Green Politics: A Modern Approach

Green Politics: A Modern Approach

Edited by Joan Bradley

Clanrye International
www.clanryeinternational.com

Clanrye International,
750 Third Avenue, 9th Floor,
New York, NY 10017, USA

ISBN: 978-1-63240-728-3

Cataloging-in-Publication Data

Green politics : a modern approach / edited by Joan Bradley.
p. cm.
Includes bibliographical references and index.
ISBN 978-1-63240-728-3
1. Green movement. 2. Environmentalism. 3. Environmental policy.
I. Bradley, Joan.
JA75.8 .G74 2018
322.44--dc23

For information on all Clanrye International publications
visit our website at www.clanryeinternational.com

CLANRYE INTERNATIONAL

Printed in China.

Contents

Preface

The political approach based on the practice of social justice, environmentalism, non-violence and grassroot democracy is known as green politics. Its main aim is to build a society which is environmentally and ecologically stable. There are many ideologies connected to this political approach namely feminism, peace movements, ecological conservation, etc. This book elucidates new techniques in green politics and their applications in a multidisciplinary approach. It provides theories and concepts used in this field in a detailed manner. Those in search of information to further their knowledge will be greatly assisted by this textbook.

A detailed account of the significant topics covered in this book is provided below:

Chapter 1- Green politics is an ideology which supports environmentally sustainable growth and social justice. The subject overlaps with topics like feminism, environmentalism, ecology and conservation. This is an introductory chapter which will introduce briefly all the significant aspects of green politics.

Chapter 2- Ecofeminism is the connection between feminism and ecology. The core belief of the subject is that patriarchy and capitalism leads to a society which causes a split between nature and culture. The chapter on green politics offers an insightful focus, keeping in mind the complex subject matter.

Chapter 3- Agrarianism is a philosophy which considers rural society to be better than urban society. It believes in promoting the simple rural life as opposed to the life people live in big cities. This chapter has been carefully written to provide an easy understanding of the varied facets of green politics.

Chapter 4- The main values and principles of green politics are pacifism, peace movement, social justice and social equality. Pacifism is the condemnation of war and violence whereas peace movement is a movement which focuses on ending war and promoting peace. This chapter elucidates the values and principles of green politics.

Chapter 5- Ecotheology is a theory that studies the relation between religion and nature. Along with the degradation of nature it also focuses on ecosystem management. The topics discussed in the chapter are of great importance to broaden the existing knowledge on green politics.

Chapter 6- Green economy mainly concentrates on decreasing environmental risks by aiming at a sustainable environment. Industries have started to adopt these practices to promote green economy. Eco-capitalism, ecological economics and deep ecology are some important topics related to the subject matter. Green politics is best understood in confluence with the major topics listed in the following chapter.

Chapter 7- The allied fields of green politics discussed in this section are food politics, climate movement, animal rights movement and environmental movement. The environmental movement is a diverse movement which centers on human rights, ecology and health. The chapter serves as a source to understand the allied fields of green politics.

It gives me an immense pleasure to thank our entire team for their efforts. Finally in the end, I would like to thank my family and colleagues who have been a great source of inspiration and support.

Editor

1

An Introduction to Green Politics

Green politics is an ideology which supports environmentally sustainable growth and social justice. The subject overlaps with topics like feminism, environmentalism, ecology and conservation. This is an introductory chapter which will introduce briefly all the significant aspects of green politics.

Green Politics

Green politics (also known as ecopolitics) is a political ideology that aims to create an ecologically sustainable society rooted in environmentalism, nonviolence, social justice, and grassroots democracy. It began taking shape in the western world in the 1970s; since then Green parties have developed and established themselves in many countries around the globe, and have achieved some electoral success.

The political term *Green* was used initially in relation to *die Grünen* (German for "the Greens"), a Green party formed in the late 1970s. The term *political ecology* is sometimes used in academic circles, but in the latter has come to represent an interdisciplinary field of study; the academic discipline offers wide-ranging studies integrating ecological social sciences with political economy in topics such as degradation and marginalization, environmental conflict, conservation and control, and environmental identities and social movements.

Supporters of green politics share many ideas with the ecology, conservation, environmentalism, feminism, and peace movements. In addition to democracy and ecological issues, green politics is concerned with civil liberties, social justice, nonviolence, sometimes variants of localism and tends to support social progressivism. The party's platform is largely considered left in the political spectrum.

The Green ideology has connections with various other ecocentric political ideologies, including ecosocialism, ecoanarchism, and ecofeminism, but to what extent these can be seen as forms of Green politics is a matter of debate.

As the left-wing 'Green' (i.e. capital 'G') political philosophy developed, there also came into separate existence unrelated and polar opposite movements on the right that include ecological components such as green conservatism and eco-capitalism.

History

Influences

Adherents to green politics tend to consider it to be part of a 'higher' worldview and not simply a political ideology. Green politics draws its ethical stance from a variety of sources, from the values of indigenous peoples, to the ethics of Gandhi, Spinoza and Uexküll. These people influenced green thought in their advocacy of long-term "seventh generation" foresight, and on the personal responsibility of every individual to make moral choices.

Of course, unease about adverse consequences of human actions on nature predates the modern concept of "environmentalism". Social commentators as far apart as ancient Rome and China complained of air, water and noise pollution.

The philosophical roots of environmentalism can be traced back to enlightenment thinkers such as Rousseau in France and, later, the author and naturalist Thoreau in America. Organised environmentalism began in late 19th Century Europe and the United States as a reaction to the Industrial Revolution with its emphasis on unbridled economic expansion.

"Green politics" first began as conservation and preservation movements, such as the Sierra Club, founded in San Francisco in 1892.

Left-green platforms of the form that make up the green parties today draw terminology from the science of ecology, and policy from environmentalism, deep ecology, feminism, pacifism, anarchism, libertarian socialism, social democracy, eco-socialism, and/or social ecology. In the 1970s, as these movements grew in influence, green politics arose as a new philosophy which synthesized their goals.

The Green Party political movement is not to be confused with the unrelated fact that in some far-right and fascist parties, nationalism has on occasion been tied into a sort of green politics which promotes environmentalism as a form of pride in the "motherland" according to a minority of authors.

German Green Party co-founder, Petra Kelly, with former German cabinet member, Otto Schily, at press conference in 1983.

Early Development

In June 1970 in the Netherlands a group called Kabouters won 5 of the 45 seats on the Amsterdam Gemeenteraad (City Council), as well as two seats each on councils in The Hague and Leeuwarden and one seat apiece in Arnhem, Alkmaar and Leiden. The Kabouters were an outgrowth of Provo's environmental White Plans and they proposed "Groene Plannen" ("Green Plans").

The first political party to be created with its basis in environmental issues was the United Tasmania Group, founded in Australia in March 1972 to fight against deforestation and the creation of a dam that would damage Lake Pedder; whilst it only gained three percent in state elections, it had, according to Derek Wall, "inspired the creation of Green parties all over the world." In May 1972, a meeting at Victoria University of Wellington, New Zealand, launched the *Values Party*, the world's first countrywide green party to contest Parliamentary seats nationally. A year later in 1973, Europe's first green party, the UK's Ecology Party, came into existence.

The German Green Party was not the first Green Party in Europe to have members elected nationally but the impression was created that they had been, because they attracted the most media attention: The German Greens, contended in their first national election in 1980. They started as a provisional coalition of civic groups and political campaigns which, together, felt their interests were not expressed by the conventional parties. After contesting the 1979 Euro elections they held a conference which identified Four Pillars of the Green Party which all groups in the original alliance could agree as the basis of a common Party platform: welding these groups together as a single Party. This statement of principles has since been utilised by many Green Parties around the world. It was this party that first coined the term "Green" ("Grün" in German) and adopted the sunflower symbol. In the 1983 federal election, the Greens won 27 seats in the Bundestag.

Further Developments

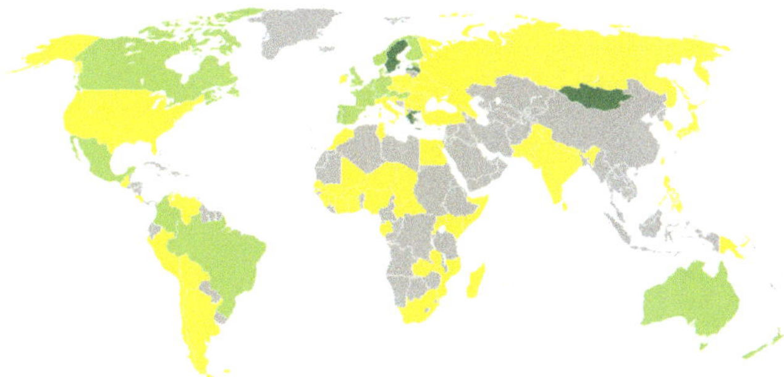

Green parties worldwide. Dark green means in government, light green in parliament and yellow in local government.

The first Canadian foray into green politics took place in the Maritimes when 11 independent candidates (including one in Montreal and one in Toronto) ran in the 1980 federal election under the banner of the Small Party. Inspired by Schumacher's Small is Beautiful, the Small Party candidates ran for the expressed purpose of putting forward an anti-nuclear platform in that election. It was not registered as an official party, but some participants in that effort went on to form the Green Party of Canada in 1983 (the Ontario Greens and British Columbia Greens were also formed that year). Current Green Party of Canada leader Elizabeth May was the instigator and one of the candidates and she was eventually elected in 2011 Canadian federal election.

In Finland, in 1995, the Green League became the first European Green Party to form part of a state-level Cabinet. The German Greens followed, forming a government with the Social Democratic Party of Germany (the "Red-Green Alliance") from 1998 to 2005. In 2001, they reached an agreement to end reliance on nuclear power in Germany, and agreed to remain in coalition and support the German government of Chancellor Gerhard Schröder in the 2001 Afghan War. This put them at odds with many Greens worldwide, but demonstrated that they were capable of difficult political tradeoffs.

In Latvia, Indulis Emsis, leader of the Green Party and part of the Union of Greens and Farmers, an alliance of a Nordic agrarian party and the Green Party, was Prime Minister of Latvia for ten months in 2004, making him the first Green politician to lead a country in the history of the world. In 2015, Emsis' party colleague, Raimonds Vējonis, was elected President of Latvia by the Latvian parliament. Vējonis became the first green head of state worldwide.

In the German state of Baden-Württenburg, the Green Party became the leader of the coalition with the Social Democrats after finishing second in the Baden-Württemberg state election, 2011. In the following state election, 2016, the Green Party became the strongest party for the first time in a German Landtag.

In 2016, the former leader of the Austrian green party (1997-2008), Alexander Van der Bellen, officially running as an independent, won the Austrian presidential election, 2016, making him the second green head of state worldwide, the first directly elected by popular vote. Van der Bellen became second in the election's first round with 21.3% of the votes, the best result for the Austrian greens in their history. He won the second round run-off against the far-right Freedom party's Norbert Hofer with 53.8% of the votes, making him the first President of Austria who was not backed by either the People's or the Social Democratic party.

Core Tenets

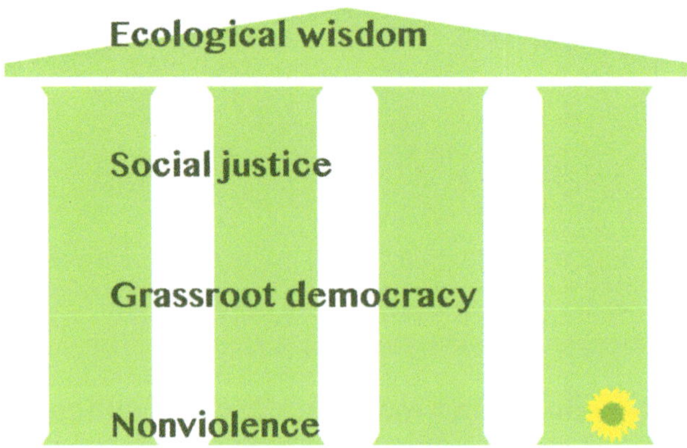

The four green pillars

According to Derek Wall, a prominent British Green proponent, there are four pillars that define Green politics:

- Ecological wisdom
- Social justice
- Grassroots democracy
- Nonviolence

In 1984, the Green Committees of Correspondence in the United States expanded the Four Pillars into Ten Key Values which, in addition to the Four Pillars mentioned above, include:

- Decentralization
- Community-based economics
- Post-patriarchal values (later translated to Feminism)
- Respect for diversity
- Global responsibility
- Future focus

In 2001, the Global Greens were organized as an international Green movement. The Global Greens Charter identified six guiding principles:

- Ecological wisdom
- Social justice
- Participatory democracy
- Nonviolence
- Sustainability
- Respect for diversity

The six guiding principles

Ecology

Economics

Green economics focuses on the importance of the health of the biosphere to human well-being. Consequently, most Greens distrust conventional capitalism, as it tends to emphasize economic growth while ignoring ecological health; the "full cost" of economic growth often includes damage to the biosphere, which is unacceptable according to green politics. Green economics considers such growth to be "uneconomic growth"— material increase that nonetheless lowers overall quality of life. Green economics inherently takes a longer term perspective than conventional economics, because such loss in quality of life is often delayed. According to green economics, the present generation should not borrow from future generations, but rather attempt to achieve what Tim Jackson calls "prosperity without growth".

Some Greens refer to productivism, consumerism and scientism as "grey", as contrasted with "green", economic views. "Grey" implies age, concrete, and lifelessness.

Therefore, adherents to green politics advocate economic policies designed to safeguard the environment. Greens want governments to stop subsidizing companies that waste resources or pollute the natural world, subsidies that Greens refer to as "dirty subsidies". Some currents of green politics place automobile and agribusiness subsidies in this category, as they may harm human health. On the contrary, Greens look to a green tax shift that are seen to encourage both producers and consumers to make ecologically friendly choices.

Green economics is on the whole anti-globalist. According to Greens, economic globalization is considered a threat to well-being, which will replace natural environments and local cultures with a single trade economy, termed the global economic monoculture.

Since green economics emphasizes biospheric health and biodiversity, an issue outside the traditional left-right spectrum, different currents within green politics incorporate ideas from socialism and capitalism. Greens on the Left are often identified as Eco-socialists, who merge ecology and environmentalism with socialism and Marxism and blame the capitalist system for environmental degradation, social injustice, inequality and conflict. Eco-capitalists, on the other hand, believe that the free market system, with some modification, is capable of addressing ecological problems. This belief is documented in the business experiences of eco-capitalists in the book, *The Gort Cloud* that describes the gort cloud as the green community that supports eco-friendly businesses.

Participatory democracy

Since the beginning, green politics has emphasized local, grassroots-level political activity and decision-making. According to its adherents, it is crucial that citizens play a direct role in the decisions that influence their lives and their environment. Therefore, green politics seeks to increase the role of deliberative democracy, based on direct citizen involvement and consensus decision making, wherever it is feasible.

Green politics also encourages political action on the individual level, such as ethical consumerism, or buying things that are made according to environmentally ethical standards. Indeed, many green parties emphasize individual and grassroots action at the local and regional levels over electoral politics. Historically, green parties have grown at the local level, gradually gaining influence and spreading to regional or provincial politics, only entering the national arena when there is a strong network of local support.

In addition, many Greens believe that governments should not levy taxes against strictly local production and trade. Some Greens advocate new ways of organizing authority to increase local control, including urban secession, bioregional democracy, and co-operative / local stakeholder ownership.

Other Issues

The sunflower is an internationally recognized symbol of Green politics.

Green politics on the whole is opposed to nuclear power and the buildup of persistent organic pollutants, supporting adherence to the precautionary principle, by which technologies are rejected unless they can be proven to not cause significant harm to the health of living things or the biosphere.

In the spirit of nonviolence, Green politics opposes the War on Terrorism and the curtailment of civil rights, focusing instead on nurturing deliberative democracy in war-torn regions and the construction of a civil society with an increased role for women.

In Europe, Green parties tend to support the creation of a democratic federal Europe.

Although Greens in the United States "call for an end to the 'War on Drugs'" and "for decriminalization of victimless crimes", they also call for developing "a firm approach to law enforcement that directly addresses violent crime, including trafficking in hard drugs".

Green platforms generally favor tariffs on fossil fuels, restricting genetically modified organisms, and protections for ecoregions or communities. In keeping with their commitment to the preservation of diversity, greens are often committed to the maintenance and protection of indigenous communities, languages, and traditions. An example of this is the Irish Green Party's commitment to the preservation of the Irish Language.

Some of the green movement has focused on divesting in fossil fuels. Academics Stand Against Poverty states "it is paradoxical for universities to remain invested in fossil fuel companies". Thomas Pogge says that the fossil fuel divestment movement can increase political pressure at events like the international climate change conference (COP). Alex Epstein of Forbes notes that it is hypocritical to ask for divestment without a boycott and that a boycott would be more effective. Some institutions that are leading by example in the academic area are Stanford University, Syracuse University, Sterling College and over 20 more. A number of cities, counties and religious institutions have also joined the movement to divest.

Organization

Local Movements

Green ideology emphasizes participatory democracy and the principle of "thinking globally, acting locally". As such, the ideal Green Party is thought to grow from the bottom up, from neighborhood to municipal to (eco-)regional to national levels. The goal is to rule by a consensus decision making process.

Strong local coalitions are considered a pre-requisite to higher-level electoral breakthroughs. Historically, the growth of Green parties has been sparked by a single issue where Greens can appeal to ordinary citizens' concerns. In Germany, for example, the Greens' early opposition to nuclear power won them their first successes in the federal elections.

Global Organization

There is a growing level of global cooperation between Green parties. Global gatherings of Green Parties now happen. The first Planetary Meeting of Greens was held 30–31 May 1992, in Rio de Janeiro, immediately preceding the United Nations Conference on Environment and Development held there. More than 200 Greens from 28 nations attended. The first formal Global Greens Gathering took place in Canberra, in 2001, with more than 800 Greens from 72 countries in attendance. The second Global Green Congress was held in São Paulo, Brazil, in May 2008, when 75 parties were represented.

Global Green networking dates back to 1990. Following the Planetary Meeting of Greens in Rio de Janeiro, a Global Green Steering Committee was created, consisting of two seats for each continent. In 1993 this Global Steering Committee met in Mexico City and authorized the creation of a Global Green Network including a Global Green Calendar, Global Green Bulletin, and Global Green Directory. The Directory was issued in several editions in the next years. In 1996, 69 Green Parties from around the world signed a common declaration opposing French nuclear testing in the South Pacific, the first statement of global greens on a current issue. A second statement was issued in December 1997, concerning the Kyoto climate change treaty.

At the 2001 Canberra Global Gathering delegates for Green Parties from 72 countries decided upon a Global Greens Charter which proposes six key principles. Over time, each Green Party can discuss this and organize itself to approve it, some by using it in the local press, some by translating it for their web site, some by incorporating it into their manifesto, some by incorporating it into their constitution. This process is taking place gradually, with online dialogue enabling parties to say where they are up to with this process.

The Gatherings also agree on organizational matters. The first Gathering voted unanimously to set up the *Global Green Network* (GGN). The GGN is composed of three representatives from each Green Party. A companion organization was set up by the same resolution: *Global Green Coordination* (GGC). This is composed of three representatives from each Federation (Africa, Europe, The Americas, Asia/Pacific). Discussion of the planned organization took place in several Green Parties prior to the Canberra meeting. The GGC communicates chiefly by email. Any agreement by it has to be by unanimity of its members. It may identify possible global campaigns to propose to Green Parties worldwide. The GGC may endorse statements by individual Green Parties. For example, it endorsed a statement by the US Green Party on the Israel-Palestine conflict.

Thirdly, Global Green Gatherings are an opportunity for informal networking, from which joint campaigning may arise. For example, a campaign to protect the New Caledonian coral reef, by getting it nominated for World Heritage Status: a joint campaign by the New Caledonia Green Party, New Caldonian indigenous leaders, the French Green Party, and the Australian Greens. Another example concerns Ingrid Betancourt, the leader of the Green Party in Colombia, the Green Oxygen Party (*Partido Verde Oxigeno*). Ingrid Betancourt and the party's Campaign Manager, Claire Rojas, were kidnapped by a hard-line faction of FARC on 7 March 2002, while travelling in FARC-controlled territory. Betancourt had spoken at the Canberra Gathering, making many friends. As a result, Green Parties all over the world have organized, pressing their governments to bring pressure to bear. For example, Green Parties in African countries, Austria, Canada, Brazil, Peru, Mexico, France, Scotland, Sweden and other countries have launched campaigns calling for Betancourt's release. Bob Brown, the leader of the Australian Greens, went to Colombia, as did an envoy from the European Federation, Alain Lipietz, who issued a report. The four Federations of Green Parties issued a message to FARC. Ingrid Betancourt was rescued by the Colombian military in Operation Jaque in 2008.

Global Green Meetings

Separately from the Global Green Gatherings, *Global Green Meetings* take place. For instance, one took place on the fringe of the World Summit on Sustainable Development in Johannesburg. Green Parties attended from Australia, Taiwan, Korea, South Africa, Mauritius, Uganda, Cameroon, Republic of Cyprus, Italy, France, Belgium, Germany, Finland, Sweden, Norway, the USA, Mexico and Chile.

The Global Green Meeting discussed the situation of Green Parties on the African continent; heard a report from Mike Feinstein, former Mayor of Santa Monica, about setting up a web site of the GGN; discussed procedures for the better working of the GGC; and decided two topics on which the Global Greens could issue statements in the near future: Iraq and the 2003 WTO meeting in Cancun.

Green Federations

The member parties of the Global Greens are organised into four continental federations:

- Federation of Green Parties of Africa
- Federation of the Green Parties of the Americas / Federación de los Partidos Verdes de las Américas
- Asia-Pacific Green Network
- European Federation of Green Parties

The European Federation of Green Parties formed itself as the European Green Party on 22 February 2004, in the run-up to European Parliament elections in June 2004, a further step in trans-national integration.

Currents

Green politics is usually said to include the green anarchism, eco-anarchism, green libertarianism, green liberalism, anti-nuclear movements, and peace movements, although these often claim not to be aligned with any party. Some claim it also includes feminism, pacifism and the animal rights movements. Some Greens support policy measures to empower women, especially mothers, great ape personhood; to oppose war, to de-escalate conflicts, and to stop proliferating technologies useful in conflict or likely to lead to conflict.

Greens on the Left adhere to eco-socialism, an ideology that combines ecology, environmentalism, socialism, and Marxism to criticize the capitalist system as the cause of ecological crises, social exclusion, inequality, and conflict. Green parties are not eco-socialist, but some Green parties around the world have or have had a significant eco-socialist membership.

Despite this stereotype, some centrist Greens may subscribe to a more classical liberal Georgist or geolibertarian philosophy emphasizing individual property rights and free-market environmentalism – and shifting taxes away from value created by labor or service and charging instead for human consumption of the wealth created by the natural world.

Greens may view the processes by which living beings compete for mates, homes, and food, ecology, and the cognitive and political sciences very differently. These differences tend to drive debate on ethics, formation of policy, and the public resolution of these differences in leadership races. There is no single "Green Ethic".

Green Party

In 1979, Daniel Brélaz became the world's first green member of a national parliament (in Switzerland).

A Green party is a formally organized political party based on the principles of green politics, such as social justice, environmentalism and nonviolence. Greens believe that these issues are inherently related to one another as a foundation for world peace. Green party platforms typically embrace social-democratic economic policies and forming coalitions with leftists. Green parties exist in nearly 90 countries around the world; many are members of Global Greens.

Definitions

There are distinctions between "green" parties and "Green" parties. Any party, faction, or politician may be labeled "green" if it emphasizes environmental causes. Indeed, the term may even be used as a verb: it is not uncommon to hear of "greening" a party or a candidate.

In contrast, formally organized Green parties may follow a coherent ideology that includes not only environmentalism, but often also other concerns such as social justice, consensus decision-making and nonviolence. Greens believe that these issues are inherently related to one another as a foundation for world peace. The best-known statement of the above Green values is the Four Pillars of the Green Party, adopted by the German Greens in 1979–1980 (but forsaken since). The Global Greens Charter lists six guiding principles which are ecological wisdom, social justice, participatory democracy, nonviolence, sustainability and respect for diversity.

Influence

Political parties campaigning on a predominantly environmental platform arose in the early 1970s in various parts of the world.

The world's first political parties to campaign on a predominantly environmental platform were the United Tasmania Group contested the April 1972 state election in Tasmania, Australia and the Values Party of New Zealand, which contested the November 1972 New Zealand general election. Their use of the name 'Green' derived from the 'Green Bans': an Australian movement of building workers who refused to build on sites of cultural and environmental significance.

The first green party in Europe was the Popular Movement for the Environment, founded in 1972 in the Swiss canton of Neuchâtel. The first national green party in Europe was PEOPLE, founded in Britain in February 1973, which eventually turned into the Ecology Party, and then the Green Party. Several other local political groups were founded in beginning of the 1970s and Fons Sprangers was probably the first Green mayor in the world, elected in 1970 in Meer, and active until 2006 for the Flemish Greens. The first political party to use the name "Green" seems to have been the Lower Saxon "Green List for Environmental Protection", founded Sept. 1, 1977.

The first Green Party to achieve national prominence was the German Green Party, famous for their opposition to nuclear power, as well as an expression of anti-centralist and pacifist values traditional to greens. They were founded in 1980 and have been in coalition governments at state level for some years. They were in federal government with the Social Democratic Party of Germany in a so-called Red-Green alliance from 1998 to 2005. In 2001, they reached an agreement to end reliance on nuclear power in Germany, and agreed to remain in coalition and support the government of Chancellor Gerhard Schröder in the 2001 Afghan War. This put them at odds with many Greens worldwide.

In Finland, in 1995, the Finnish Green Party was the first European Green party to be part of a national Cabinet. Other Green parties that have participated in government at national level include the Groen! (formerly Agalev) and Ecolo in Belgium, Les Verts in France and the Green Party in Ireland. In the Netherlands GroenLinks ("GreenLeft") was founded in 1990 from four small left-wing parties and is now a stable faction in the Dutch parliament.

Around the world, individuals have formed many Green parties over the last thirty years. Green parties now exist in most countries with democratic systems: from Canada to Peru; from Norway to South Africa; from Ireland to Mongolia. There is Green representation at national, regional and local levels in many countries around the world.

Most of the Green parties are formed to win elections, and so organize themselves by the presented electoral or political districts. But that does not apply universally: The Green Party of Alaska is organized along bioregional lines to practice bioregional democracy.

Alliances

Depending on local conditions or issues, platforms and alliances may vary. In line with the goal of democracy, neighboring ecoregions may require different policies or protections.

Green parties are often formed in a given jurisdiction by a coalition of scientific ecologists, community environmentalists, and local (or national) leftist groups or groups concerned with peace or citizens rights.

A Red-Green alliance is an alliance between Green parties and social democratic parties. Such alliances are typically formed for the purpose of elections (mostly in first-past-the-post election systems), or, after elections, for the purpose of forming a government.

Some Greens, such as those in Hawaii, find more effective alliances with more conservative groups (Blue-Green alliance) or indigenous peoples who seek to prevent disruption of traditional ways of life or to save ecological resources they depend on. Although Greens find much to support in fostering these types of alliances with groups of historically different backgrounds, they also feel strongly about forming diverse communities through encouragement of diversity in social and economic demographics in communities, especially in the United States.

Alliances often highlight strategic differences between participating in parties and advancing the values of the Green movement. For example, Greens became allied with centre-right parties to oust the centre-left ruling Institutional Revolutionary Party of Mexico. Ralph Nader, the 2000 presidential nominee of the U.S. Greens, campaigned with ultra-conservative Pat Buchanan on joint issues such as farm policy and bans on corporate funding of election campaigns, although this "alliance" between Nader and Buchanan was very specifically limited to the purpose of showing that there was broad support for certain specific issues, across the political spectrum.

U.S. Greens grew dramatically throughout 2001. However, stable coalitions (such as that in Germany) tend to be formed between elections with left-wing parties on social issues, and 'the grassroots right' on such issues as irresponsible corporate subsidies and public ethics.

On 13 June 2007, the Irish Green Party, represented by six members of parliament or TDs, agreed to go into a coalition government for the first time in their history, with Fianna Fáil. Fianna Fáil and the Green Party were joined by the Progressive Democrats in a coalition also supported by a number of independent members of parliament (the Progressive Democrats later dissolved as a party, though their members remained in parliament). The Green Party held two Cabinet seats, as well as two junior ministries, until their withdrawal from government in January 2011. They lost all of their six seats in the following month's general election, but regained two in the 2016 General Election.

In the Czech Republic, the Green Party was part of the governing coalition, together with the conservative Civic Democratic Party (ODS) and the Christian Democrats (KDU–ČSL) from January 2007 until the government collapsed in March 2009. The Finnish and French Green parties are now the only ones actively participating in government.

Green Parties

Belgian and German Roots

The first green parties in Europe were founded in the late 1970s, following the rise of environmental awareness and the development of new social movements. Green parties in Belgium first made a breakthrough. Belgium had Green members of parliament elected first in the 1970s, and with seats on the local council, held the balance of power in the city of Liege, so were the first to go into coalition with the ruling party on that council. In 1979 political campaigns and dissident groups feeling underrepresented in west German politics formed a coalition to contest the 1979 elections to the European Parliament.

Although they did not win any seats, the groups in this association formally agreed to become a party and won a breakthrough in the German national elections of 1983. They were not the first Green Party in Europe to have members elected nationally but the impression was created that they had been, because they attracted the most media attention. This was partly due to their charismatic leader Petra Kelly, a German who was of interest to the American media because she had an American step-father. Since its foundation in 1980 and merger with Alliance 90 after the German reunification, Kelly's party, now named Alliance '90/The Greens, has become one of Europe's most important Green parties. It played an important role in the formation of national-level Green parties in other countries such as Spain as well. The forerunner of the Green Party in the United Kingdom was the PEOPLE Party, formed in Coventry in 1972. It changed its name to the Ecology Party in 1973 and the Green Party in 1985.

1984–1989: A New Political Force

In 1984 Greens agreed a common platform for the European Parliament Elections and the first Green Members of the European Parliament were elected here. Germany, a stronghold of the Green movement, elected seven MEPs; two more came from Belgium and two from the Netherlands. As those nine MEPs did not entitle the Greens to form a parliamentary group on their own, they concluded an alliance with MEPs from Italy, Denmark, and regionalists from Flanders and Ireland to form the GRAEL (Green Alternative European Link) group, also known as the Rainbow Group. Politically they engaged in the fight against environmental pollution, nuclear energy (1986 saw the Chernobyl disaster), the promotion of animal protection and the campaign against the demolition of Brussels by speculation fuelled by the presence of the European institutions.

Since the 1990s

Indulis Emsis of the Latvian Green Party became Prime Minister of Latvia in 2004, and the world's first green head of government.

After years of co-operation between the national Green parties, they formed a pan-European alliance that unites most European Green parties. The Greens are a party within the European parliament with 46 seats, as of June 2009. It has a long-standing alliance with the European Free

Alliance (EFA), an alliance of "stateless nations", such as the Welsh nationalist Plaid Cymru and Scottish National Party. Together European Green Party/EFA have 58 seats and they are the fourth largest party in the European Parliament.

While on many issues European Greens practice the same policies, one issue divides European Green parties: the European Union. Some Green parties, like the Dutch GreenLeft, the Green Party of England and Wales, the Swiss Green Party, the Irish Green Party and the German Alliance '90/The Greens, are pro-European while some, like the Green party in Sweden, are moderately eurosceptic.

Some Green parties have been part of governing coalitions. The first one was the Finnish Green League that entered government in 1995. The Italian Federation of the Greens, the French Greens, the German Alliance '90/The Greens and both Belgian Green parties, the French-speaking Ecolo and the Dutch-speaking Agalev were part of government during the late 1990s. Most successful was the Latvian Green Party, who supplied the Prime Minister of Latvia in 2004. The Swedish Green party was a long term supporter of the social-democratic minority government until the election 2006 when the social-democratic party lost. The Irish Green Party were in government from 2007 until January 2011 when the party withdrew their support for the ruling coalition. During their period in office, the Irish Green Party held two Cabinet portfolios including Minister for the Environment, Heritage and Local Government and Minister for Communications, Energy and Natural Resources.

Raimonds Vējonis, President of Latvia

Alexander Van der Bellen, President of Austria

Vējonis (elected 2015) and Van der Bellen (elected 2016) are Europe's first two green heads of state.

In Scandinavia, left-wing socialist parties have formed the Nordic Green Left Alliance. These parties have the same ideals as European Greens. However, they do not cooperate with the Global Greens or European Greens, but instead form a combined parliamentary group with the Party of the European Left, which unites communists and post-communists. There is one exception, in 2004 the MEP for Danish Socialist People's Party has left the Nordic Green Left parliamentary group and has joined the Green parliamentary group in the European parliament. The Socialist People's Party is currently an observer at the European Green Party and the Global Greens. Outside of Scandinavia, in 2004, Latvia became the first country in the world to have a Green politician become Head of Government, but in 2006 the Green Party received only 16.71 percent of the vote. In the Estonia 2007 parliamentary elections, the Estonian Greens won 7.1 percent of the vote, and a mandate for six seats in the country's parliament, the Riigikogu. Other significant electoral results for European green parties include Germany's Alliance '90/The Greens in the 2002 federal election and France's Europe Ecology – The Greens in the 2012 legislative election, in which they both won more seats than any other third party.

British MP Caroline Lucas is the world's first green politician elected under a first-past-the-post system.

In some countries Greens have found it difficult to win any representation in the national parliament. Three reasons can be found for this. It includes countries with a first past the post electoral system, such as the United Kingdom. However, despite the first past the post system in the United Kingdom, the Green Party of England and Wales won their first seat in the House of Commons when Caroline Lucas won the seat of Brighton Pavilion in 2010. The Scottish Green Party has had success in the devolved Scottish Parliament and the Irish Green Party in Northern Ireland has had success in the devolved Northern Ireland Assembly, where the first-past-the-post system is not used. In countries where a party with similar ideals is stronger, such as Norway and Denmark, Green parties tend to perform worse. In some Eastern European countries, like Romania and Poland, Green parties are still in the process of formation and have therefore not gained enough support. The Green Party of Bulgaria is a part of the left-wing Coalition for Bulgaria, currently in opposition. It has no parliamentary representation but it did supply one Deputy Minister in the government of Prime Minister Sergey Stanishev from 2005 to 2009.

The European Green Party has worked to support weak Green parties in European countries. Until recently, they were giving support to Green parties in the Mediterranean countries. These Green parties are now making electoral gains, e.g. in Spain, Greece and Republic of Cyprus, or getting organized to do so, e.g. in Malta. Therefore, the EGP is now turning its attention to Eastern Europe –

all these countries have Green parties, but in materially poor Eastern Europe the success of Green Parties is very patchy, except for Hungary, where the local Green party, Politics Can Be Different (LMP), has succeeded in getting into the parliament and many city councils.

Asia and Oceania

Green parties have achieved national or state parliamentary representation in New Zealand, Australia and Vanuatu. In New Zealand the Green Party of New Zealand currently holds 14 seats in the New Zealand House of Representatives after the 2011 general election. The Australian Greens hold 10 seats in the Australian Senate and one seat in the Australian House of Representatives. Since 2004, they have received more votes than any other third party in every federal election. They also have representation in the upper and lower houses of state parliaments of five states and in the unicameral chamber of one territory. Greens also hold representative positions in local government across New Zealand and Australia (where a number of local government authorities are controlled by Green councilors). The Greens took the seat of Melbourne from the Australian Labor Party in 2010 with candidate Adam Bandt. This is the first time the Greens have won a Lower House seat at a general election (although they have previously won two seats at by-elections).

Proportional representation in the Australian Senate and in New Zealand has strengthened the position of the Australian Greens and the Green Party of New Zealand and enabled them to participate directly in legislatures and policy-making committees. In countries following British-style 'first past the post' electoral rules, Green parties face barriers to gaining federal or provincial/regional/state seats. The Australian Labor Party's practice of allocating a portion of ALP ticket votes to Australian Greens has helped bring AG candidates into parliament.

In the 2008 ACT election in Australia, the Greens won 15.6% of the vote, winning 4 out of 17 seats. Shane Rattenbury was elected the speaker of the assembly, the first time a Green party member had held such a position in any parliament or assembly in Australia. However, they retained only one seat at the 2012 election in the same territory.

The Green Confederation (*Confédération Verte*) in Vanuatu won 3 out of 52 seats in the last general election in October 2012. Its most prominent member is Moana Carcasses Kalosil, who became Prime Minister in March 2013. Carcasses, a Green liberal, does not lead a Green government, but a broad coalition government in which he is the only Green minister.

There is a Papua New Guinea Greens Party, but it does not have any members in Parliament. There was briefly a Green Party in Fiji from 2008 to 2013; as Parliament was suspended by the military regime during this time, the party was not able to take part in any election before all parties were deregistered in 2013.

North America

As of the 41st general election in Canada, held on May 2, 2011, there was only one federally elected member of the Green Party of Canada in the Canadian House of Commons, its leader Elizabeth May. However, at the dissolution of the 41st Parliament on August 2, 2015, the Green Party of Canada held two seats in the House of Commons, the second seat belonging to formerly Independent MP Bruce Hyer who was elected to the House of Commons as a member of the NDP in 2011. Only May won re-election to the 42nd parliament. There remains no federal representation by the Green Party of the United States in the U.S. Congress. Accordingly, in these countries, Green parties focus on electoral reform. In Mexico, however, the *Partido Verde Ecologista* has 17 deputies and four senators in Congress as a result of the 2006 elections.

Canada

The first Green parties in Canada (both federal and provincial) were founded in 1983. The strongest provincial Green parties are the Green Party of British Columbia and the Green Party of Ontario; although in Ontario they have yet to win a seat in provincial election. As of May 2013, the first ever BC elected green member was elected, and in 2014 a green MLA was elected to the New Brunswick legislature. Federally, the Green Party of Canada received 3.91% of the popular vote in the 2011 federal election and its support and influence continues to rise. Support waned slightly in the 2011 federal election when it captured 3.91% of the popular vote, down from the 6.8% captured in the 2008 federal election; support fell another 0.46% in the 2015 federal election. Although Ms. May was the first elected Member of Parliament, the first seat was gained in the Canadian House of Commons on August 30, 2008 when sitting Independent MP Blair Wilson joined the party. May continues to sit as a Member of Parliament for Saanich-Gulf Islands and has exerted a considerable amount of legislative influence on a number of issues ranging from denying unanimous consent for military intervention in Libya, playing a central role in exposing changes to environmental legislation hidden in the omnibus Budget Bill C-38, to introducing a private member's bill to develop a national strategy on Lyme disease.

In the 2008 Vancouver municipal election, Stuart Mackinnon, a member of the Vancouver Green Party, was elected to the Vancouver Parks Board. Since that time former Green Party of British Columbia leader, and deputy-leader of the federal Green party, Adriane Carr won the Greens' first seat on Vancouver City Council, in 2011 municipal elections.

United States

In the United States, Greens first ran for public office in 1985. Since then, the Green Party of the United States has claimed electoral victories at the municipal, county and state levels. The first U.S. Greens to be elected were David Conley and Frank Koehn in Wisconsin 1986. Each was elected to a position on the County Board of Supervisors in Douglas and Bayfield counties respectively. Keiko Bonk was first elected in 1992 in Hawaii County, becoming Official Chairwoman in 1995. The first Green Party mayor was Kelly Weaverling, elected in Cordova, AK in 1991.

Approximately 100 Greens hold elected office across the US as of October 2016. The first US Green elected to a state legislature was Audie Bock in 1999, to the California State Assembly, followed by John Eder to the Maine House of Representatives in 2002 and 2004 and Richard Carroll to the Arkansas House of Representatives in 2008. While in office in 2003 in the New Jersey General Assembly, incumbent Matt Ahearn made a party switch to Green for the remainder of his term.

The Green Party has contested six U.S. presidential elections: in 1996 and 2000 with Ralph Nader for President and Winona LaDuke as Vice President, in 2004 with David Cobb for President and Pat LaMarche for Vice President, and in 2008 with Cynthia McKinney for President and Rosa Clemente for Vice President. In 2000, Nader received more votes for president than any Green

Party candidate before or since. Jill Stein ran for president on the Green ticket in 2012 and 2016; the vice-presidential candidates were Cheri Honkala in 2012 and Ajamu Baraka in 2016. Stein, who received over one million votes in the 2016 race, went on to lead the 2016 election recounts in several states.

Developing World

Green parties in the developing world are often organized with help from those in other nations. As of 2002 the foundation of Green parties has been the most notable in Africa.

Other than hosting the first Afghanistan peace conference as part of the German government, Green parties in the developed world have made few concrete moves to spread their values using the diplomatic channels. This is usually seen as one of the responsibilities of the Green movement - allowing parties to concentrate on their voters. However, the leader of the Kenyan Green Party, Wangari Maathai, won the Nobel Peace Prize in 2004, enhancing the image of Green parties across the third world.

In the greater Middle East region, a few Green parties have been created, such as the Green Party of Pakistan since 2002 and Green Party of India since 2010, and the Green Party of Saudi Arabia, but many of these Green parties are underground organizations.

Brazil

Marina Silva's Green Party in the Brazilian presidential election, 2010 won 19.33% of the vote of the first round (more than any other third party), taking enough votes from Dilma Rousseff of the incumbent PT party to stop her getting the 50% needed to avoid a second round of which she went on to beat Jose Serra of the opposition PSDB party. Despite only winning numerically the vote in only 1 state; the relatively electorally small Federal District which holds the capital of Brazil; the Green Party interestingly came second to Dilma in Amapa, Amazonas, Pernambuco and in Rio de Janeiro in front of Jose Serra. The Green party also tied with Serra in Ceara, both having 16.36% of the vote, although Dilma won the state. Maranhão also had a 1.5% difference in vote between Serra and Marina with the Green Party coming third. In São Paulo, the Green Party numerically had their strongest showing with almost 5 million votes taking 20.77% of the vote. Success in the legislature however only amounted to winning 2 more seats to a total of 15 in the Chamber of Deputies and the loss of their only senate seat.

Colombia

In the 1990s, the Oxygen Green Party was created under the leadership of Ingrid Betancourt but dissolved after her infamous kidnapping. Later, the Visionaries Party was created by Antanas Mockus whose ideals earned him the Bogotá Mayoral Office twice. In the 2010 Colombian presidential election a green party has been created under the name Colombian Green Party, with former Bogotá mayor Antanas Mockus being the leader.

Lebanon

The Green Party of Lebanon was founded in 2008 as a secular party. Its first president was Philippe Skaff, CEO of Grey Advertising. The party debuted with the May 2010 municipal elections. In 2011, the party became the first political party in Lebanon to elect a female leader when Nada Zaarour was elected its president .

Turkey

Greens and the Left Party of the Future is a left-libertarian and green party in Turkey. It was founded on 25 November 2012 as a merger of the Greens Party and the Equality and Democracy Party.

Prominent members include Murat Belge, left-liberal political author and columnist for *Taraf*; Kutluğ Ataman, filmmaker and contemporary artist; and Ufuk Uras, former Istanbul deputy and president of the Freedom and Solidarity Party.

The party is one of the participants in the Peoples' Democratic Congress, a political initiative instrumental in founding the Peoples' Democratic Party in 2012.

2

Fundamental Concepts of Green Politics

Ecofeminism is the connection between feminism and ecology. The core belief of the subject is that patriarchy and capitalism leads to a society which causes a split between nature and culture. The chapter on green politics offers an insightful focus, keeping in mind the complex subject matter.

Ecofeminism

Ecofeminism is a term that links feminism with ecology. Its advocates say that paternalistic/capitalistic society has led to a harmful split between nature and culture. Early ecofeminists propagated that the split can only be healed by the feminine instinct for nurture and holistic knowledge of nature's processes. Modern ecofeminism, or feminist ecocriticism, eschews such essentialism and instead focuses more on intersectional questions, such as how the nature-culture split enables the oppression of female and nonhuman bodies. It is also an activist and academic movement that sees critical connections between the exploitation of nature and the domination over women both caused by men.

Definition

Ecofeminism describes movements and philosophies that link feminism with ecology. This movement seeks to eradicate all forms of social injustice, not just injustice against women and the environment. The term is believed to have been coined by the French writer Françoise d'Eaubonne in her book *Le Féminisme ou la Mort* (1974). From arguments that there are particular and significant connections between women and nature, ecofeminism relates the oppression and domination of all subordinate groups (women, people of color, children, the poor) to the oppression and domination of nature (animals, land, water, air, etc.). All of these subordinate groups have been subject to oppression, domination, exploitation, and colonization from the Western patriarchal society that emphasizes and values men. Ecofeminists believe that these connections are illustrated through traditionally «feminine» values such as reciprocity, nurturing and cooperation, which are present both among women and in nature.

Several feminists make the distinction that it is not because women are women or "feminine" that they relate to nature, but because of their similar states of oppression by the same male-dominant force. The marginalization is evident in the gendered language used to describe nature and the animalized language used to describe women. Some discourses link women specifically to the environment because of their traditional social role as a nurturer and caregiver.

As there are several different types of feminism and different beliefs held by feminists, there are different versions of ecofeminism. Charlene Spretnak has argued that there are three main 'paths' to ecofeminism, and that they draw in a diverse group of people, which strengthens any movement, such as: 1) through the study of political theory as well as history; 2) through the belief and study of nature-based religions; 3) through environmentalism.

In the 1993 essay entitled "Ecofeminism: Toward Global Justice and Planetary Health" authors Greta Gaard and Lori Gruen outline what they call the "ecofeminist framework." The essay provides a wealth of data and statistics in addition to laying out the theoretical aspects of the ecofeminist critique. The framework described is intended to establish ways of viewing and understanding our current global situations so that we are better able to understand how we arrived at this point and what may be done to ameliorate the ills. The four sides of the frame are: the mechanistic materialist model of the universe that resulted from the scientific revolution and the subsequent reduction of all things into mere resources to be optimized, dead inert matter to be used; the rise of patriarchal religions and their establishment of gender hierarchies along with their denial of immanent divinity; self and other dualisms and the inherent power and domination ethic it entails; and capitalism and its intrinsic need for the exploitation, destruction and instrumentalization of animals, earth and people for the sole purpose of creating wealth. They hold that these four factors have brought us to what ecofeminists see as a "separation between nature and culture" that is the root source of our planetary ills.

Vandana Shiva says that women have a special connection to the environment through their daily interactions and this connection has been ignored. She says that women in subsistence economies who produce "wealth in partnership with nature, have been experts in their own right of holistic and ecological knowledge of nature's processes." She makes the point that "these alternative modes of knowing, which are oriented to the social benefits and sustenance needs are not recognized by the capitalist reductionist paradigm, because it fails to perceive the interconnectedness of nature, or the connection of women's lives, work and knowledge with the creation of wealth." Shiva blames this failure on the West's patriarchy, and the patriarchal idea of what development is. According to Shiva, patriarchy has labeled women, nature, and other groups not growing the economy as "unproductive".

History

Background

Ecofeminism was coined as a term in the 1970s. Women participated in the environmental movements, specifically preservation and conservation, much earlier than this. Beginning in the late 20th century, women worked in efforts to protect wildlife, food, air and water. These efforts depended largely on new developments in the environmental movement from influential writers, such as Henry David Thoreau, Aldo Leopold, John Muir, and Rachel Carson. Fundamental examples of women's efforts in the 20th century are the books *Silent Spring* by Rachel Carson and *Refuge* by Terry Tempest Williams. These works truly opened American's eyes to the environmental harm they were perpetuating, and created a platform for change.

Ecofeminist author Karren Warren lists Aldo Leopold's essay "Land Ethic" (1949) as a fundamental work to the ecofeminist conception, as Leopold was the first to pen an ethic for the land which understands all non-human parts of that community (animals, plants, land, air, water) as equal to and in a relationship with humans. This inclusive understanding of the environment launched the modern preservation movement and illustrated how issues can be viewed through a framework of caring.

Susan A. Mann an eco-feminist and professor of sociological and feminist theory considers the roles women played in these activisms to be the starter for ecofeminism in later centuries. Mann associates the beginning of ecofeminism not with feminists but with women of different race and class backgrounds who made connections among gender, race, class and environmental issues. This ideal is upheld through the notion that in activist and theory circles marginalized groups must be included in the discussion. In early environmental and women's movements, issues of varying races and classes were often separated.

Movements of the 1970s

In northern India in 1973, women took part in the Chipko movement to protect forests from deforestation. They clung to the trees in an act of peaceful protest so that loggers could not cut them down. These women demonstrated the importance of the environment to them, and were leaders as well as activists.

In Kenya in 1977, the Green Belt Movement was initiated by Professor Wangari Maathai, environmental and political activist, and is ongoing today. It is rural tree planting program led by women, which Maathai designed to help prevent desertification in the area. The program created a 'green belt' of at least 1,000 trees around villages, and gives participants the ability to take charge in their communities. In later years, the Green Belt Movement was an advocate for informing and empowering citizens through seminars for civic and environmental education, as well as holding national leaders accountable for their actions and instilling agency in citizens.

In 1978 in New York, mother and environmentalist Lois Gibbs led her community in protest after discovering that their entire neighborhood, Love Canal, was built on top of a toxic dump site. The toxins in the ground were causing illness among children and reproductive issues among women, as well as birth defects in babies born to pregnant women exposed to the toxins. The Love Canal movement eventually led to the evacuation and relocation of nearly 800 families by the federal government.

1980s and 1990s

After the beginning of the environmental movement in the early 1970s intersections among feminists and other social justice movements emerged. The feminists that took interests in these movements explored how oppressions were linked through gender, race, class and ecology, as well as species and ideas of nationhood. They took cues from Green Party politics, peace movements, and direct action movements. These feminists developed texts, such as *Women and Nature* (Susan Griffin 1978), *The Death of Nature* (Carolyn Merchant 1980) and *Gyn/Ecology* (Mary Daly 1978). These texts helped to propel the association between domination by man on women and the domination of culture on nature. From these texts feminist activism of the 1980s linked ideas of ecology and the environment. For example, conferences for women devoted to living on the earth and protests against nuclear testing and other militarism that oppresses femininity.

In 1980 and 1981, members of such a conference organized an peaceful protest at the pentagon. Women stood, hand in hand, demanding equal rights (including social, economic, and reproductive rights) as well as an end to militaristic actions taken by the government and exploitation of the community (people and the environment). This movement is known as the Women's Pentagon Actions.

In 1985, the Akwesasne Mother's Milk Project was launched by Katsi Cook. This study was funded by the government, and investigated how the higher level of contaminants in water near the Mohawk reservation impacted babies. It revealed that through breast milk, Mohawk children were being exposed to 200% more toxins than children not on the reservation. Toxins contaminate water all over the world, but to due environmental racism, certain subversive groups are exposed to a much higher amount.

The Greening of Harlem Coalition is another example of an ecofeminist movement. In 1989, Bernadette Cozart founded the coalition, which is responsible for many urban gardens around Harlem. Cozart's goal is to turn vacant lots into community gardens. This is economically beneficial, and also provides a way for very urban communities to be in touch with nature and each other. The majority of people interested in this project (as noted in 1990) were women. Through these gardens, they were able to participate in and become leaders of their communities. Urban greening exists in other places as well. Beginning in 1994, a group of African-American women in Detroit have developed city gardens, and call themselves the Gardening Angels. Similar garden movements have been occurring globally.

The development of vegetarian ecofeminism can be traced to the mid-80s and 90s, where it first appeared in writing. However, the roots of a vegetarian ecofeminist view can be traced back further by looking at sympathy for non-humans and counterculture movements of the 1960s and 1970s. At the culmination of the decade ecofeminism had spread to both coasts and articulated an intersectional analysis of women and the environment. Eventually, challenging ideas of environmental classism and racism, resisting toxic dumping and other threats to the impoverished.

In the 1980s and 1990s some began to see the advancing theories in ecofeminism as essentialist. Through analysis done by post structural and third wave feminists it was argued that ecofeminism equated women with nature. This dichotomy is dangerous because it groups all women into one category and enforces the very societal norms that feminism is trying to break. Out of this critique rose the anti-essentialist argument. Ecofeminist and author Noel Sturgeon says in an interview that what anti-essentialists are critiquing is a strategy used to mobilize large and diverse groups of both theorists and activists.

1990s–present

Coming out of the 90s, ecofeminism met a lot of criticism from anti-essentialist feminism, which heavily critiqued what they viewed as essentialism. The essentialist view saw ecofeminism as reinforcing and growing patriarchal dominance and norms. Feminist thoughts surrounding ecofeminism grew in some areas as it was criticized; vegetarian ecofeminism contributed intersectional analysis; and ecofeminisms that analyzed animal rights, labor rights and activisms as they could draw lines among oppressed groups. To some, the inclusion of non-human animals also became to be viewed as essentialist. According to ecofeminist and author Charlene Spretnak, modern ecofeminism is concerned about a variety of issues, including reproductive technology, equal pay and equal rights, toxic poisoning, Third World development, and more.

Ecofeminism as it propelled into the 21st century became aware of the criticisms, and in response ecofeminists with a materialist lens began doing research and renaming the topic, i.e. queer ecologies, global feminist environmental justice, and gender and the environment.

Major critiques

The major criticism of ecofeminism is that it is essentialist. The ascribed essentialism appears in two main areas:

- Adherence to strict dichotomy between men and women: Some eco-feminist critiques are that the dichotomy between women and men and nature and culture creates a dualism that is too stringent and focused in the difference of women and men. That eco-feminism too strongly correlates the social status of women with the social status of nature, rather than the non-essentialist view that women along with nature both have masculine and feminine qualities, and that just like feminine qualities have often been seen as less worthy, nature is also seen as having lesser value than culture, or the qualities involved in these concepts.

- Divergent view regarding participation in oppressive structures: As opposed to radical and liberation-based feminist movements, mainstream feminism which is most tightly bound with hegemonic social status strives to promote equality within the existing social and political structure, such as making it possible for women to occupy positions of power in business, industry and politics, using direct involvement as the main tactic for achieving pay equity and influence. In contrast, many ecofeminists oppose active engagement in these areas, as these are the very structures that the movement intends to dismantle.

Social ecologist and feminist Janet Biehl has criticized ecofeminism for focusing too much on a mystical connection between women and nature and not enough on the actual conditions of women.

Rosemary Radford Ruether also critiques this focus on mysticism over work that focuses on helping women, but argues that spirituality and activism can be combined effectively in ecofeminism.

Concepts

In *Ecofeminism* (1993) authors Vandana Shiva, Maria Mies and Evan Bondi ponder modern science and its acceptance as a universal and value-free system. Instead, they view the dominant stream of modern science as a projection of Western men's values. The privilege of determining what is considered scientific knowledge has been controlled by men, and for the most part of history restricted to men. Bondi and Miles list examples including the medicalization of childbirth and the industrialization of plant reproduction.

Bondi argues that the medicalization of childbirth has marginalized midwife knowledge and changed the natural process of childbirth into a procedure dependent on specialized technologies and appropriated expertise. A common claim within ecofeminist literature is that patriarchal structures justify their dominance through binary opposition, these include but are not limited to: heaven/earth, mind/body, male/female, human/animal, spirit/matter, culture/nature and white/non-white. Oppression is reinforced by assuming truth in these binaries and instilling them as 'marvelous to behold' through religious and scientific constructs.

The application of ecofeminism to animal rights has established vegetarian ecofeminism, which asserts that "omitting the oppression of animals from feminist and ecofeminist analyses ... is inconsistent with the activist and philosophical foundations of both feminism (as a "movement to end all forms of oppression") and ecofeminism." It puts into practice "the personal is political" for it believes that "meat-eating is a form of patriarchal domination...that suggests a link between male violence and a meat-based diet." Vegetarian ecofeminism combines sympathy with the analysis of culture and politics to refine a system of ethics and action.

Ecofeminism as materialist is another common theme in ecofeminism. A materialist view connects some institutions such as labor, power and property as the source of domination over women and nature. There are connections made between these subjects because similarly there are varying values in production and reproduction.

Spiritual ecofeminism is another branch of ecofeminism, and is popular among ecofeminist authors such as Starhawk, Riane Eisler, Carol J. Adams, and more. Starhawk calls this an earth-based spirituality, which recognizes that the Earth is alive, that we are interconnected, as well as a community. Spiritual ecofeminism is not linked to one specific religion, but is centered around values of caring, compassion, and non-violence. Often, ecofeminists refer to more ancient traditions, such as the worship of Gaia, the Goddess of nature and spirituality (also known as Mother Earth).

Theorists

- Judi Bari – Bari was a member of the Earth First! movement and says she was targeted due to her womanhood.
- Françoise d'Eaubonne – Called upon women to lead an ecological revolution in order to save the planet. This entailed revolutionizing gender relations and human relations with the natural world.
- Greta Gaard – Greta Gaard is an American ecofeminist scholar and activist. Her major contributions to the field connect ideas of queer theory, vegetarianism, and animal liberation. Her major theories include ecocriticism which works to include literary criticism and composition to inform ecofeminism and other feminist theories to address wider range of social issues within ecofeminism. She is an ecological activist and leader in the U.S. Green Party, and the Green Movement.
- Sallie McFague – A prominent ecofeminist theologian, McFague uses the metaphor of God's body to represent the universe at large. This metaphor values inclusive, mutualistic and interdependent relations amongst all things.
- Carolyn Merchant – Historian of science who taught at Berkeley for many years. Her book *The Death of Nature: Women, Ecology and the Scientific Revolution* is a classic ecofeminist text.
- Mary Mellor – UK sociologist who moved to ecofeminist ideas from an interest in cooperatives. Her books - *Breaking the Boundaries* and *Feminism and Ecology* are grounded in a materialist analysis.
- Maria Mies – Mies is a German social critic who has been involved in feminist work

throughout Europe and India. She works particularly on the intersections of patriarchy, poverty, and the environment on a local and global scale.

- Val Plumwood – Val Plumwood, formerly Val Routley, was an Australian ecofeminist intellectual and activist, who was prominent in the development of radical ecosophy from the early 1970s through the remainder of the 20th century. In her works "Feminism and the Mastery of Nature" she describes the relationship of mankind and the environment relating to an eco-feminist ideology.
- Rosemary Radford Ruether – Has written 36 books and over 600 articles exploring the intersections of feminism, theology, and creation care.
- Ariel Salleh – Australian ecofeminist with a global perspective; a founding editor of the journal *Capitalism Nature Socialism*; author of two books and some 200 articles examining links with deep and social ecology, green politics and eco-socialism.
- Vandana Shiva – Shiva is a physicist, author, activist, feminist and philosopher from India. She was a participant in the Chipko movement of the 1970s, which used non-violent activism to protest and prevent deforestation in the Garhwal Himalayas of Uttarakhand, India then in Uttar Pradesh.
- Charlene Spretnak – Spretnak is an American writer largely known for her writing on ecology, politics and spirituality. Through these writings Spretnak has become a prominent ecofeminist. She has written many books which discuss ecological issues in terms of effects with social criticisms, including feminism. Spretnak works had a major influence in the development of the Green Party. She has also won awards based on her visions on ecology and social issues as well as feminist thinking.
- Starhawk – An American writer and activist Starhawk is known for her work in spiritualism and ecofeminism. She advocates for social justice in issues surrounding nature and spirit. These social justice issues fall under the scope of feminism and ecofeminism. She believes in fighting oppression through intersectionality and the importance of spirituality, eco consciousness and sexual and gender liberation.
- Douglas Vakoch – An American ecocritic whose edited volumes include *Ecofeminism and Rhetoric: Critical Perspectives on Sex, Technology, and Discourse* (2011), *Feminist Ecocriticism: Environment, Women, and Literature* (2012), and (with Sam Mickey) *Women and Nature?: Beyond Dualism in Gender, Body, and Environment* (2017).
- Karen Warren – received her B.A. in philosophy from the University of Minnesota (1970) and her Ph.D. from the University of Massachusetts-Amherst in 1978. Before her long tenure at Macalester College, which began in 1985, Warren was Professor of Philosophy at St. Olaf College in the early 1980s. Warren was the Ecofeminist-Scholar-in-Residence at Murdoch University in Australia. In 2003, she served as an Oxford University Round Table Scholar and as Women's Chair in Humanistic Studies at Marquette University in 2004. She has spoken widely on environmental issues, feminism, critical thinking skills and peace studies in many international locations including Buenos Aires, Gothenburg, Helsinki, Oslo, Manitoba, Melbourne, Moscow, Perth, the U.N. Earth Summit in Rio de Janeiro (1992), and San Jose.

Eco-socialism

Eco-socialism, green socialism or socialist ecology is an ideology merging aspects of socialism with that of green politics, ecology and alter-globalization or anti-globalization. Eco-socialists generally believe that the expansion of the capitalist system is the cause of social exclusion, poverty, war and environmental degradation through globalization and imperialism, under the supervision of repressive states and transnational structures.

Eco-socialists advocate dismantling capitalism, focusing on common ownership of the means of production by freely associated producers, and restoring the commons. Caroline Lucas, leader of the Green Party of England and Wales, has described her party's brand of socialism as appealing to both middle-class environmentalists as well as working-class socialists.

Ideology

Eco-socialists are critical of many past and existing forms of both Green politics and socialism. They are often described as "Red Greens" - adherents to Green politics with clear anti-capitalist views, often inspired by Marxism (Red Greens are in contrast to eco-capitalists and Green anarchists).

The term "watermelon" is commonly applied, often pejoratively, to Greens who seem to put "social justice" goals above ecological ones, implying they are "green on the outside but red on the inside"; the term is usually attributed to either Petr Beckmann or, more frequently, Warren T. Brookes, both critics of environmentalism, and is common in Australia, New Zealand and the United States.

A New Zealand website, *The Watermelon*, uses the term proudly, stating that it is "green on the outside and liberal on the inside", while also citing "socialist political leanings", reflecting the use of the term "liberal" to describe the left wing in many English-speaking countries. Red Greens are often considered "fundies" or "fundamentalist greens", a term usually associated with Deep Ecology even though the German Green Party "fundi" faction included eco-socialists, and eco-socialists in other Green Parties, like Derek Wall, have been described in the press as fundies.

Eco-socialists also criticise bureaucratic and elite theories of self-described socialism such as Maoism, Stalinism and what other critics have termed bureaucratic collectivism or state capitalism. Instead, eco-socialists focus on imbuing socialism with ecology while keeping the emancipatory goals of "first-epoch" socialism. Eco-socialists aim for communal ownership of the means of production by "freely associated producers" with all forms of domination eclipsed, especially gender inequality and racism.

This often includes the restoration of commons land in opposition to private property, in which local control of resources valorizes the Marxist concept of use value above exchange value. Practically, eco-socialists have generated various strategies to mobilise action on an internationalist basis, developing networks of grassroots individuals and groups that can radically transform society through nonviolent "prefigurative projects" for a post-capitalist, post-statist world.

History

1880S–1930S – Marx, Morris and Influence on The Russian Revolution

Contrary to the depiction of Karl Marx by some environmentalists, social ecologists and fellow socialists as a productivist who favoured the domination of nature, eco-socialists have revisited Marx's writings and believe that he "was a main originator of the ecological world-view". Eco-socialist authors, like John Bellamy Foster and Paul Burkett, point to Marx's discussion of a "metabolic rift" between man and nature, his statement that "private ownership of the globe by single individuals will appear quite absurd as private ownership of one man by another" and his observation that a society must "hand it [the planet] down to succeeding generations in an improved condition". Nonetheless, other eco-socialists feel that Marx overlooked a "recognition of nature in and for itself", ignoring its "receptivity" and treating nature as "subjected to labor from the start" in an "entirely active relationship".

William Morris, the English novelist, poet and designer, is largely credited with developing key principles of what was later called eco-socialism. During the 1880s and 1890s, Morris promoted his eco-socialist ideas within the Social Democratic Federation and Socialist League.

Following the Russian Revolution, some environmentalists and environmental scientists attempted to integrate ecological consciousness into Bolshevism, although many such people were later purged from the Communist Party of the Soviet Union. The "pre-revolutionary environmental movement", encouraged by revolutionary scientist Aleksandr Bogdanov and the Proletkul't organisation, made efforts to "integrate production with natural laws and limits" in the first decade of Soviet rule, before Joseph Stalin attacked ecologists and the science of ecology and the Soviet Union fell into the pseudo-science of the state biologist Trofim Lysenko, who "set about to rearrange the Russian map" in ignorance of environmental limits.

Ecoanarchism

Green anarchism, or ecoanarchism, is a school of thought within anarchism which puts a particular emphasis on environmental issues. An important early influence was the thought of the American anarchist Henry David Thoreau and his book *Walden* as well as Leo Tolstoy and Elisee Reclus. In the late 19th century there emerged anarcho-naturism as the fusion of anarchism and naturist philosophies within individualist anarchist circles in France, Spain, Cuba and Portugal. Several anarchists from the mid-20th century, including Herbert Read, Ethel Mannin, Leopold Kohr, Jacques Ellul, and Paul Goodman, also held proto-environmental views linked to their anarchism. Mannin's 1944 book *Bread and Roses: A Utopian Survey and Blue-Print* has been described by anarchist historian Robert Graham as setting forth "an ecological vision in opposition to the prevailing and destructive industrial organization of society". Important contemporary currents are anarcho-primitivism and social ecology.

Social Ecology and Communalism

Social ecology is closely related to the work and ideas of Murray Bookchin and influenced by anarchist Peter Kropotkin. Social ecologists assert that the present ecological crisis has its roots in human social problems, and that the domination of human-over-nature stems from the domination of human-over-human. In 1958, Murray Bookchin defined himself as an anarchist, seeing parallels between anarchism and ecology. His first book, *Our Synthetic Environment,* was published under the pseudonym Lewis Herber in 1962, a few months before Rachel Carson's *Silent Spring*. The book described a broad range of environmental ills but received little attention because of its political radicalism. His groundbreaking essay "Ecology and Revolutionary Thought" introduced ecology as a concept in radical politics. In 1968 he founded another group that published the influential *Anarchos* magazine, which published that and other innovative essays on post-scarcity and on ecological technologies such as solar and wind energy, and on decentralization and miniaturization. Lecturing throughout the United States, he helped popularize the concept of ecology to the counterculture.

Murray Bookchin

Post-Scarcity Anarchism is a collection of essays written by Murray Bookchin and first published in 1971 by Ramparts Press. It outlines the possible form anarchism might take under conditions of post-scarcity. It is one of Bookchin's major works, and its radical thesis provoked controversy for being utopian and messianic in its faith in the liberatory potential of technology. Bookchin argues that post-industrial societies are also post-scarcity societies, and can thus imagine "the fulfillment of the social and cultural potentialities latent in a technology of abundance". The self-administration of society is now made possible by technological advancement and, when technology is used in an ecologically sensitive manner, the revolutionary potential of society will be much changed. In 1982, his book *The Ecology of Freedom* had a profound impact on the emerging ecology movement, both in the United States and abroad. He was a principal figure in the Burlington Greens in 1986-90, an ecology group that ran candidates for city council on a program to create neighborhood democracy.

Bookchin later developed a political philosophy to complement social ecology which he called "Communalism" (spelled with a capital "C" to differentiate it from other forms of communalism). While originally conceived as a form of Social anarchism, he later developed Communalism into a separate ideology which incorporates what he saw as the most beneficial elements of Anarchism, Marxism, syndicalism, and radical ecology.

Politically, Communalists advocate a network of directly democratic citizens' assemblies in individual communities/cities organized in a confederal fashion. This method used to achieve this is called Libertarian Municipalism which involves the establishment of face-to-face democratic institutions which are to grow and expand confederally with the goal of eventually replacing the nation-state.

1970s–1990s – Rise of Environmentalism and Engagement with Marxism and 'Actually Existing Socialism'

In the 1970s, Barry Commoner, suggesting a left-wing response to the *Limits to Growth* model that predicted catastrophic resource depletion and spurred environmentalism, postulated that capitalist technologies were chiefly responsible for environmental degradation, as opposed to population pressures. East German dissident writer and activist Rudolf Bahro published two books addressing the relationship between socialism and ecology – *The Alternative in Eastern Europe* and *Socialism and Survival* – which promoted a 'new party' and led to his arrest, for which he gained international notoriety.

At around the same time, Alan Roberts, an Australian Marxist, posited that people's unfulfilled needs fuelled consumerism. Fellow Australian Ted Trainer further called upon socialists to develop a system that met human needs, in contrast to the capitalist system of created wants. A key development in the 1980s was the creation of the journal *Capitalism, Nature, Socialism* (CNS) with James O'Connor as founding editor and the first issue in 1988. The debates ensued led to a host of theoretical works by O'Connor, Carolyn Merchant, Paul Burkett and others.

The Australian Democratic Socialist Party launched the Green Left Weekly newspaper in 1991, following a period of working within Green Alliance and Green Party groups in formation. This ceased when the Australian Greens adopted a policy of proscription of other political groups in August 1991. The DSP also published a comprehensive policy resolution, "Socialism and Human Survival" in book form in 1990, with an expanded second edition in 1999 entitled "Environment, Capitalism & Socialism".

1990s Onwards – Engagement with the Anti-globalization Movement and The Ecosocialist Manifesto

The 1990s saw the socialist feminists Mary Mellor and Ariel Salleh address environmental issues within an eco-socialist paradigm. With the rising profile of the anti-globalization movement in the Global South, an "environmentalism of the poor", combining ecological awareness and social justice, has also become prominent. David Pepper also released his important work, *Ecosocialism: From Deep Ecology to Social Justice*, in 1994, which critiques the current approach of many within Green politics, particularly deep ecologists.

In 2001, Joel Kovel, a social scientist, psychiatrist and former candidate for the Green Party of the United States (GPUS) Presidential nomination in 2000, and Michael Löwy, an anthropologist and member of the Reunified Fourth International (a principal Trotskyist organisation), released *An ecosocialist manifesto*, which has been adopted by some organisations and suggests possible routes for the growth of eco-socialist consciousness. Kovel's 2002 work, *The Enemy of Nature: The End of Capitalism or the End of the World?*, is considered by many to be the most up-to-date exposition of eco-socialist thought.

In October 2007, the International Ecosocialist Network was founded in Paris.

Influence on Current Green and Socialist Movements

Currently, many Green Parties around the world, such as the Dutch Green Left Party (Groen-Links) , contain strong eco-socialist elements. Radical Red-green alliances have been formed in many countries by eco-socialists, radical Greens and other radical left groups. In Denmark, the Red-Green Alliance was formed as a coalition of numerous radical parties. Within the European Parliament, a number of far-left parties from Northern Europe have organized themselves into the Nordic Green Left Alliance. Red Greens feature heavily in the Green Party of Saskatchewan (in Canada but not necessarily affiliated to the Green Party of Canada) and then-recently GPUS that officially adopted ideology within party.

The Green Party of England and Wales features an eco-socialist group, Green Left, that was founded in June 2005 and whose members hold a number of influential positions within the party, including both the former Principal Speakers Siân Berry and Dr. Derek Wall, himself an eco-socialist and Marxist academic, as well as prominent Green Party candidate and human rights activist Peter Tatchell. Many Marxist organisations also contain eco-socialists, as evidenced by Löwy's involvement in the reunified Fourth International and *Socialist Resistance*, a British Marxist newspaper that reports on eco-socialist issues and has published two collections of essays on eco-socialist thought: *Ecosocialism or Barbarism?*, edited by Jane Kelly and Sheila Malone, and *The Global Fight for Climate Justice*, edited by Ian Angus with a foreword by Derek Wall.

Influence on "Existing Socialist" Regimes

Eco-socialism has had a minor influence over developments in the environmental policies of what can be called "existing socialist" regimes, notably the People's Republic of China. Pan Yue, Deputy Director of the PRC's State Environmental Protection Administration, has acknowledged the influence of eco-socialist theory on his championing of environmentalism within China, which has gained him international acclaim (including being nominated for the Person of the Year Award 2006 by *The New Statesman*, a British current affairs magazine). Yue stated in an interview that, while he often finds eco-socialist theory "too idealistic" and lacking "ways of solving actual problems", he believes that it provides "political reference for China's scientific view of development", "gives socialist ideology room to expand" and offers "a theoretical basis for the establishment of fair international rules" on the environment.

He echoes much of eco-socialist thought, attacking international "environmental inequality", refusing to focus on technological fixes and arguing for the construction of "a harmonious, resource-saving and environmentally-friendly society". He also shows a knowledge of eco-socialist history, from the convergence of radical green politics and socialism and their political "red-green alliances" in the post-Soviet era. This focus on eco-socialism has informed an essay, *On Socialist Ecological Civilisation*, published in September 2006, which, according to chinadialogue, "sparked debate" in China. The current Constitution of Bolivia, promulgated in 2009, is the first both ecologic and pro-socialist Constitution in the world, making the Bolivian state officially ecosocialist.

Ecosocialist International Network (EIN) and other International Eco-socialist Organisations

In 2007, it was announced that attempts to form an Ecosocialist International Network (EIN) would be made and an inaugural meeting of the International occurred on October 7, 2007 in Paris. The meeting attracted "more than 60 activists from Argentina, Australia, Belgium, Brazil, Canada, Cyprus, Denmark, France, Greece, Italy, Switzerland, United Kingdom, and the United States" and elected a Steering Committee featuring representatives from Britain, the United States, Canada, France, Greece, Argentina, Brazil and Australia, including Joel Kovel, Michael Löwy, Derek Wall, Ian Angus (editor of Climate and Capitalism in Canada) and Ariel Salleh. The Committee states that it wants "to incorporate members from China, India, Africa, Oceania and Eastern Europe". EIN held its second international conference in January 2009, in association with the next World Social Forum in Brazil". The conference released The Belem Ecosocialist Declaration.

International networking by eco-socialists has already been seen in the Praxis Research and Education Center, a group on international researchers and activists. Based in Moscow and established in 1997, Praxis, as well as publishing books "by libertarian socialists, Marxist humanists, anarchists, [and] syndicalists", running the Victor Serge Library and opposing war in Chechnya, states that it believes "that capitalism has brought life on the planet near to the brink of catastrophe, and that a form of ecosocialism needs to emerge to replace capitalism before it is too late".

Critique of Capitalist Expansion and Globalisation

Merging aspects of Marxism, socialism, environmentalism and ecology, eco-socialists generally believe that the capitalist system is the cause of social exclusion, inequality and environmental degradation through globalization and imperialism under the supervision of repressive states and transnational structures.

In the *Ecosocialist manifesto*, Kovel and Löwy suggest that capitalist expansion causes both "crises of ecology" through "rampant industrialization" and "societal breakdown" that springs "from the form of imperialism known as globalization". They believe that capitalism's expansion "exposes ecosystems" to pollutants, habitat destruction and resource depletion, "reducing the sensuous vitality of nature to the cold exchangeability required for the accumulation of capital", while submerging "the majority of the world's people to a mere reservoir of labor power" as it penetrates communities through "consumerism and depoliticization".

Other eco-socialists, like Wall, highlight how, in the Global South, free-market capitalism structures economies to produce export-geared crops that take water from traditional subsistence farms, increasing hunger and the likelihood of famine; furthermore, forests are increasingly cleared and enclosed to produce cash crops that separate people from their local means of production and aggravate poverty. Wall shows that many of the world's poor have access to the means of production through "non-monetised communal means of production", such as subsistence farming, but, despite providing for need and a level of prosperity, these are not included in conventional economics measures, like GNP.

Wall therefore views neo-liberal globalization as "part of the long struggle of the state and commercial interests to steal from those who subsist" by removing "access to the resources that sustain ordinary people across the globe". Furthermore, Kovel sees neoliberalism as "a return to the pure logic of capital" that "has effectively swept away measures which had inhibited capital's aggressivity, replacing them with naked exploitation of humanity and nature"; for Kovel, this "tearing down of boundaries and limits to accumulation is known as globalization", which was "a deliberate response to a serious accumulation crisis (in the 1970s) that had convinced the leaders of the global economy to install what we know as neoliberalism.".

Furthermore, Guha and Martinez-Alier blame globalization for creating increased levels of waste and pollution, and then dumping the waste on the most vulnerable in society, particularly those in the Global South. Others have also noted that capitalism disproportionately affects the poorest in the Global North as well, leading to examples of resistance such as the environmental justice movement in the United States, consisting of working-class people and ethnic minorities who highlight the tendency for waste dumps, major road projects and incinerators to be constructed around socially excluded areas. However, as Wall highlights, such campaigns are often ignored or persecuted precisely because they originate among the most marginalized in society: the African-American radical green religious group MOVE, campaigning for ecological revolution and animal rights from Philadelphia, had many members imprisoned or even killed by US authorities from the 1970s onwards.

Eco-socialism disagrees with the elite theories of capitalism, which tend to label a specific class or social group as conspirators who construct a system that satisfies their greed and personal desires. Instead, eco-socialists suggest that the very system itself is self-perpetuating, fuelled by "extra-human" or "impersonal" forces. Kovel uses the Bhopal industrial disaster as an example. Many anti-corporate observers would blame the avarice of those at the top of many multi-national corporations, such as the Union Carbide Corporation in Bhopal, for seemingly isolated industrial accidents. Conversely, Kovel suggests that Union Carbide were experiencing a decrease in sales that led to falling profits, which, due to stock market conditions, translated into a drop in share values. The depreciation of share value made many shareholders sell their stock, weakening the company and leading to cost-cutting measures that eroded the safety procedures and mechanisms at the Bhopal site. Though this did not, in Kovel's mind, make the Bhopal disaster inevitable, he believes that it illustrates the effect market forces can have on increasing the likelihood of ecological and social problems.

Use and Exchange Value

Eco-socialism focuses closely on Marx's theories about the contradiction between use values and exchange values. Kovel posits that, within a market economy, goods are not produced to meet needs but are produced to be exchanged for money that we then use to acquire other goods; as we have to keep selling in order to keep buying, we must persuade others to buy our goods just to ensure our survival, which leads to the production of goods with no previous use that can be sold to sustain our ability to buy other goods.

Such goods, in an eco-socialist analysis, produce exchange values but have no use value. Eco-socialists like Kovel stress that this contradiction has reached a destructive extent, where certain essential activities - such as caring for relatives full-time and basic subsistence – are unrewarded, while unnecessary commodities earn individuals huge fortunes and fuel consumerism and resource depletion.

The "Second Contradiction" of Capitalism

James O'Connor argues for a "second contradiction" of underproduction, to complement Marx's "first" contradiction of capital and labor. While the second contradiction is often considered a theory of environmental degradation, O'Connor's theory in fact goes much further. Building on the work of Karl Polanyi, along with Marx, O'Connor argues that capitalism necessarily undermines the "conditions of production" necessary to sustain the endless accumulation of capital. These conditions of production include soil, water, energy, and so forth. But they also include an adequate public education system, transportation infrastructures, and other services that are not produced directly by capital, but which capital needs in order accumulate effectively. As the conditions of production are exhausted, the costs of production for capital increase. For this reason, the second contradiction generates an underproduction crisis tendency, with the rising cost of inputs and labor, to complement the overproduction tendency of too many commodities for too few customers. Like Marx's contradiction of capital and labor, the second contradiction therefore threatens the system's existence.

In addition, O'Connor believes that, in order to remedy environmental contradictions, the capitalist system innovates new technologies that overcome existing problems but introduce new ones.

O'Connor cites nuclear power as an example, which he sees as a form of producing energy that is advertised as an alternative to carbon-intensive, non-renewable fossil fuels, but creates long-term radioactive waste and other dangers to health and security. While O'Connor believes that capitalism is capable of spreading out its economic supports so widely that it can afford to destroy one ecosystem before moving onto another, he and many other eco-socialists now fear that, with the onset of globalization, the system is running out of new ecosystems. Kovel adds that capitalist firms have to continue to extract profit through a combination of intensive or extensive exploitation and selling to new markets, meaning that capitalism must grow indefinitely to exist, which he thinks is impossible on a planet of finite resources.

The Role of The State and Transnational Organisations

Capitalist expansion is seen by eco-socialists as being "hand in glove" with "corrupt and subservient client states" that repress dissent against the system, governed by international organisations "under the overall supervision of the Western powers and the superpower United States", which subordinate peripheral nations economically and militarily. Kovel further claims that capitalism itself spurs conflict and, ultimately, war. Kovel states that the 'War on Terror', between Islamist extremists and the United States, is caused by "oil imperialism", whereby the capitalist nations require control over sources of energy, especially oil, which are necessary to continue intensive industrial growth - in the quest for control of such resources, Kovel argues that the capitalist nations, specifically the United States, have come into conflict with the predominantly Muslim nations where oil is often found.

Eco-socialists believe that state or self-regulation of markets does not solve the crisis "because to do so requires setting limits upon accumulation", which is "unacceptable" for a growth-orientated system; they believe that terrorism and revolutionary impulses cannot be tackled properly "because to do so would mean abandoning the logic of empire". Instead, eco-socialists feel that increasing repressive counter-terrorism increases alienation and causes further terrorism and believe that state counter-terrorist methods are, in Kovel and Löwy's words, "evolving into a new and malignant variation of fascism". They echo Rosa Luxemburg's "stark choice" between "socialism or barbarism", which was believed to be a prediction of the coming of fascism and further forms of destructive capitalism at the beginning of the twentieth century (Luxemburg was in fact murdered by proto-fascist Freikorps in the revolutionary atmosphere of Germany in 1919).

Tensions Within the Eco-Socialist Discourse

Reflecting tensions within the environmental and socialist movements, there is some conflict of ideas. In practice however, a synthesis is emerging which calls for democratic regulation of industry in the interests of people and the environment, nationalisation of some key (environmental) industries, local democracy and an extension of co-ops and the library principle.

Critique of other Forms of Green Politics

Eco-socialists criticise many within the Green movement for not being overtly anti-capitalist, for working within the existing capitalist, statist system, for voluntarism, or for reliance on technological fixes. The eco-socialist ideology is based on a critique of other forms of Green politics, including various forms of Green economics, Localism, Deep Ecology, Bioregionalism and even some manifestations of radical green ideologies such as Eco-feminism and Social Ecology.

As Kovel puts it, eco-socialism differs from Green politics at the most fundamental level because the 'Four Pillars' of Green politics (and the 'Ten Key Values' of the US Green Party) do not include the demand for the emancipation of labour and the end of the separation between producers and the means of production. Many eco-socialists also oppose Malthusianism and are alarmed by the gulf between Green politics in the Global North and the Global South.

Opposition to Within-System Approaches, Voluntarism and Technological Fixes

Eco-socialists are highly critical of those Greens who favour "working within the system". While eco-socialists like Kovel recognise the ability of within-system approaches to raise awareness, and believe that "the struggle for an ecologically rational world must include a struggle for the state", he believes that the mainstream Green movement is too easily co-opted by the current powerful socio-political forces as it "passes from citizen-based activism to ponderous bureaucracies scuffling for 'a seat at the table'".

For Kovel, capitalism is "happy to enlist" the Green movement for "convenience", "control over popular dissent" and "rationalization". He further attacks within-system green initiatives like carbon trading, which he sees as a "capitalist shell game" that turns pollution "into a fresh source of profit". Brian Tokar has further criticised carbon trading in this way, suggesting that it augments existing class inequality and gives the "largest 'players'... substantial control over the whole 'game'".

In addition, Kovel criticises the "defeatism" of voluntarism in some local forms of environmentalism that do not connect: he suggests that they can be "drawn off into individualism" or co-opted to the demands of capitalism, as in the case of certain recycling projects, where citizens are "induced to provide free labor" to waste management industries who are involved in the "capitalization of nature". He labels the notion on voluntarism "ecopolitics without struggle".

Technological fixes to ecological problems are also rejected by eco-socialists. Saral Sarkar has updated the thesis of 1970s 'limits to growth' to exemplify the limits of new capitalist technologies such as hydrogen fuel cells, which require large amounts of energy to split molecules to obtain hydrogen. Furthermore, Kovel notes that "events in nature are reciprocal and multi-determined" and can therefore not be predictably "fixed"; socially, technologies cannot solve social problems because they are not "mechanical". He posits an eco-socialist analysis, developed from Marx, that patterns of production and social organisation are more important than the forms of technology used within a given configuration of society.

Under capitalism, he suggests that technology "has been the *sine qua non* of growth" - thus he believes that, even in a world with hypothetical "free energy", the effect would be to lower the cost of automobile production, leading to the massive overproduction of vehicles, "collapsing infrastructure", chronic resource depletion and the "paving over" of the "remainder of nature". In the modern world, Kovel considers the supposed efficiency of new post-industrial commodities is a "plain illusion", as miniaturized components involve many substances and are therefore non-recyclable (and, theoretically, only simple substances could be retrieved by burning out-of-date equipment, releasing more pollutants). He is quick to warn "environmental liberals" against over-selling the virtues of renewable energies that cannot meet the mass energy consumption of the era; although he would still support renewable energy projects, he believes it is more important to restructure societies to reduce energy use before relying on renewable energy technologies alone.

Critique of Green Economics

Eco-socialists have based their ideas for political strategy on a critique of several different trends in Green economics. At the most fundamental level, eco-socialists reject what Kovel calls "ecological economics" or the "ecological wing of mainstream economics" for being "uninterested in social transformation". He furthers rejects the Neo-Smithian school, who believe in Adam Smith's vision of "a capitalism of small producers, freely exchanging with each other", which is self-regulating and competitive.

The school is represented by thinkers like David Korten who believe in "regulated markets" checked by government and civil society but, for Kovel, they do not provide a critique of the expansive nature of capitalism away from localised production and ignore "questions of class, gender or any other category of domination". Kovel also criticises their "fairy-tale" view of history, which refers to the abuse of "natural capital" by the materialism of the Scientific Revolution, an assumption that, in Kovel's eyes, seems to suggest that "nature had toiled to put the gift of capital into human hands", rather than capitalism being a product of social relations in human history.

Other forms of Community-based economics are also rejected by eco-socialists such as Kovel, including followers of E. F. Schumacher and some members of the Cooperative movement, for advocating "no more than a very halting and isolated first step". He thinks that their principles are "only partially realizable within the institutions of cooperatives in capitalist society" because "the internal cooperation" of cooperatives is "forever hemmed in and compromised" by the need to expand value and compete within the market. Marx also believed that cooperatives within capitalism make workers into "their own capitalist by enabling them to use the means of production for the employment of their own labour".

For Kovel and other eco-socialists, Community-based economics and Green Localism are "a fantasy" because "strict localism belongs to the aboriginal stages of society" and would be an "ecological nightmare at present population levels" due to "heat losses from a multitude of dispersed sites, the squandering of scarce resources, the needless reproduction of effort, and cultural impoverishment". While he feels that small-scale production units are "an essential part of the path towards an ecological society", he sees them not as "an end in itself"; in his view, small enterprises can be either capitalist or socialist in their configuration and therefore must be "consistently anti-capitalist", through recognition and support of the emancipation of labour, and exist "in a dialectic with the whole of things", as human society will need large-scale projects, such as transport infrastructures.

He highlights the work of steady-state theorist Herman Daly, who exemplifies what eco-socialists see as the good and bad points of ecological economics — while Daly offers a critique of capitalism and a desire for "workers ownership", he only believes in workers ownership "kept firmly within a capitalist market", ignoring the eco-socialist desire for struggle in the emancipation of labour and hoping that the interests of labour and management today can be improved so that they are "in harmony".

Critique of Deep Ecology

Despite the inclusion of both in political factions like the 'Fundies' of the German Green Party, eco-socialists and deep ecologists hold markedly opposite views. Eco-socialists like Kovel have attacked deep ecology because, like other forms of Green politics and Green economics, it features "virtuous souls" who have "no internal connection with the critique of capitalism and the emancipation of labor". Kovel is particularly scathing about deep ecology and its "fatuous pronouncement" that Green politics is "neither left nor right, but ahead", which, for him, ignores the notion that "that which does not confront the system comes its instrument".

Even more scathingly, Kovel suggests that in "its effort to decentre humanity within nature", deep ecologists can "go too far" and argue for the "splitting away of unwanted people", as evidenced by their desire to preserve wilderness by removing the groups that have lived there "from time immemorial". Kovel thinks that this lends legitimacy to "capitalist elites", like the US State Department and the World Bank, who can make preservation of wilderness a part of their projects that "have added value as sites for ecotourism" but remove people from their land. Between 1986 and 1996, Kovel notes that over three million people were displaced by "conservation projects"; in the making of the US National Parks, three hundred Shoshone Indians were killed in the development of Yosemite.

Kovel believes that deep ecology has affected the rest of the Green movement and led to calls from restrictions on immigration, "often allying with reactionaries in a... cryptically racist quest". Indeed, he finds traces of deep ecology in the "biological reduction" of Nazism, an ideology many "organicist thinkers" have found appealing, including Herbert Gruhl, a founder of the German Green Party (who subsequently left when it became more Left-wing) and originator of the phrase "neither left nor right, but ahead". Kovel warns that, while 'ecofascism' is confined to a narrow band of far right intellectuals and disaffected white power skinheads who involved themselves alongside far left groups in the anti-globalization movement, it may be "imposed as a revolution from above to install an authoritarian regime in order to preserve the main workings of the system" in times of crisis.

Critique of Bioregionalism

Bioregionalism, a philosophy developed by writers like Kirkpatrick Sale who believe in the self-sufficiency of "appropriate bioregional boundaries" drawn up by inhabitants of "an area", has been thoroughly critiqued by Kovel, who fears that the "vagueness" of the area will lead to conflict and further boundaries between communities. While Sale cites the bioregional living of Native Americans, Kovel notes that such ideas are impossible to translate to populations of modern proportions, and evidences the fact that Native Americans held land in commons, rather than private property - thus, for eco-socialists, bioregionalism provides no understanding of what is needed to transform society, and what the inevitable "response of the capitalist state" would be to people constructing bioregionalism.

Kovel also attacks the problems of self-sufficiency. Where Sale believes in self-sufficient regions "each developing the energy of its peculiar ecology", such as "wood in the northwest [USA]", Kovel asks "how on earth" these can be made sufficient for regional needs, and notes the environmental damage of converting Seattle into a "forest-destroying and smoke-spewing wood-burning" city. Kovel also questions Sale's insistence on bioregions that do "not require connections with the outside, but within strict limits", and whether this precludes journeys to visit family members and other forms of travel.

Critique of Variants of Eco-feminism

Like many variants of socialism and Green politics, eco-socialists recognise the importance of "the gendered bifurcation of nature" and support the emancipation of gender as it "is at the root of patriarchy and class". Nevertheless, while Kovel believes that "any path out of capitalism must also be eco-feminist", he criticises types of ecofeminism that are not anti-capitalist and can "essentialize women's closeness to nature and build from there, submerging history into nature", becoming more at place in the "comforts of the New Age Growth Centre". These limitations, for Kovel, "keep ecofeminism from becoming a coherent social movement".

Critique of Social Ecology

While having much in common with the radical tradition of Social Ecology, eco-socialists still see themselves as distinct. Kovel believes this is because Social Ecologists see hierarchy "in-itself" as the cause of ecological destruction, whereas eco-socialists focus on gender and class domination embodied in capitalism and recognise that forms of authority that are not "an expropriation of human power for self-aggrandizement", such as a student-teacher relationship that is "reciprocal and mutual", are beneficial.

In practice, Kovel describes Social Ecology as continuing the anarchist tradition of non-violent direct action, which is "necessary" but "not sufficient" because "it leaves unspoken the question of building an ecological society beyond capital". Furthermore, Social Ecologists and anarchists tend to focus on the state alone, rather than the class relations behind state domination (in the view of Marxists). Kovel fears that this is political, springing from historic hostility to Marxism among anarchists and sectarianism, which he points out as a fault of the "brilliant" but "dogmatic" founder of Social Ecology, Murray Bookchin.

Opposition to Malthusianism and Neo-Malthusianism

While Malthusianism and eco-socialism overlap within the Green movement because both address over-industrialism, and despite the fact that Eco-socialists, like many within the Green movement, are described as neo-Malthusian because of their criticism of economic growth, Eco-socialists are opposed to Malthusianism. This divergence stems from the difference between Marxist and Malthusian examinations of social injustice - whereas Marx blames inequality on class injustice, Malthus argued that the working-class remained poor because of their greater fertility and birth rates.

Neo-Malthusians have slightly modified this analysis by increasing their focus on overconsumption - nonetheless, eco-socialists find this attention inadequate. They point to the fact that Malthus did not thoroughly examine ecology and that Garrett Hardin, a key Neo-Malthusian, suggested that further enclosed and privatised land, as opposed to commons, would solve the chief environmental problem, which Hardin labeled the 'Tragedy of the Commons'.

The "Two Varieties of Environmentalism"

Guha and Martinez-Alier attack the gulf between what they see as the two "varieties of environmentalism" - the environmentalism of the North, an aesthetic environmentalism that is the privilege of wealthy people who no longer have basic material concerns, and the environmentalism of the South, where people's local environment is a source of communal wealth and such issues are a question of survival. Nonetheless, other eco-socialists, such as Wall, have also pointed out that capitalism disproportionately affects the poorest in the Global North as well, leading to examples of resistance such as the environmental justice movement in the US and groups like MOVE.

Critique of Other Forms of Socialism

Eco-socialists choose to use the term 'socialist', despite "the failings of its twentieth century interpretations", because it "still stands for the supersession of capital" and thus "the name, and the reality" must "become adequate for this time". Eco-socialists have nonetheless often diverged with other Marxist movements. Eco-socialism has also been partly influenced by and associated with agrarian socialism as well as some forms of Christian socialism, especially in the United States.

Critique of 'Actually Existing Socialisms'

For Kovel and Lowy, eco-socialism is "the realization of the "first-epoch" socialisms" by resurrecting the notion of "free development of all producers", distancing themselves from "the attenuated, reformist aims of social democracy and the productivist structures of the bureaucratic variations of socialism", such as forms of Leninism and Stalinism. They ground the failure of past socialist movements in "underdevelopment in the context of hostility by existing capitalist powers", which led to "the denial of internal democracy" and "emulation of capitalist productivism". Kovel believes that the forms of 'actually existing socialism' consisted of "public ownership of the means of production", rather than meeting "the true definition" of socialism as "a free association of producers", with the Party-State bureaucracy acting as the "alienating substitute 'public'".

In analysing the Russian Revolution, Kovel feels that "conspiratorial" revolutionary movements "cut off from the development of society" will "find society an inert mass requiring leadership from above". From this, he notes that the anti-democratic Tsarist heritage meant that the Bolsheviks, who were aided into power by World War One, were a minority who, when faced with a counter-revolution and invading Western powers, continued "the extraordinary needs of 'war communism'", which "put the seal of authoritarianism" on the revolution; thus, for Kovel, Lenin and Trotsky "resorted to terror", shut down the Soviets (workers' councils) and emulated "capitalist efficiency and productivism as a means of survival", setting the stage for Stalinism.

Lenin, in Kovel's eyes, came to oppose the nascent Bolshevik environmentalism and its champion Aleksandr Bogdanov, who was later attacked for "idealism"; Kovel describes Lenin's philosophy as "a sharply dualistic materialism, rather similar to the Cartesian separation of matter and consciousness, and perfectly tooled to the active working over of the dead, dull matter by the human hand", which led him to want to overcome Russian backwardness through rapid industrialization. This tendency was, according to Kovel, augmented by a desire to catch-up with the West and the "severe crisis" of the revolution's first years.

Furthermore, Kovel quotes Trotsky, who believed in a Communist "superman" who would "learn how to move rivers and mountains". Kovel believes that, in Stalin's "revolution from above" and mass terror in response to the early 1930s economic crisis, Trotsky's writings "were given official imprimatur", despite the fact that Trotsky himself was eventually purged, as Stalinism attacked "the very notion of ecology... in addition to ecologies". Kovel adds that Stalin "would win the gold medal for enmity to nature", and that, in the face of massive environmental degradation, the inflexible Soviet bureaucracy became increasingly inefficient and unable to emulate capitalist accumulation, leading to a "vicious cycle" that led to its collapse.

Critique of the Wider Socialist Movement

Beyond the forms of 'actually existing socialism', Kovel criticises socialists in general as treating ecology "as an afterthought" and holding "a naive faith in the ecological capacities of a working-class defined by generations of capitalist production". He exemplifies David McNally, who advocated increasing consumption levels under socialism, which, for Kovel, contradicts any notion of natural limits. He also criticises McNally's belief in releasing the "positive side of capital's self-expansion" after the emancipation of labor; instead, Kovel argues that a socialist society would "seek not to become larger" but would rather become "more *realized*", choosing sufficiency and eschewing economic growth. Kovel further adds that the socialist movement was historically conditioned by its origins in the era of industrialization so that, when modern socialists like McNally advocate a socialism that "cannot be at the expense of the range of human satisfaction", they fail "to recognize that these satisfactions can be problematic with respect to nature when they have been historically shaped by the domination of nature".

Eco-socialist Strategy

Eco-socialists generally advocate the non-violent dismantling of capitalism and the state, focusing on collective ownership of the means of production by freely associated producers and restoration of the Commons. To get to an eco-socialist society, eco-socialists advocate working-class anti-capitalist resistance but also believe that there is potential for agency in autonomous, grassroots individuals and groups across the world who can build "prefigurative" projects for non-violent radical social change.

These prefigurative steps go "beyond the market and the state" and base production on the enhancement of use values, leading to the internationalization of resistance communities in an 'Eco-socialist Party' or network of grassroots groups focused on non-violent, radical social transformation. An 'Eco-socialist revolution' is then carried out.

Agency

Many eco-socialists, like Alan Roberts, have encouraged working-class action and resistance, such as the 'green ban' movement in which workers refuse to participate in projects that are ecologically harmful. Similarly, Kovel focuses on working-class involvement in the formation of eco-socialist parties or their increased involvement in existing Green Parties; however, he believes that, unlike many other forms of socialist analysis, "there is no privileged agent" or revolutionary class, and that there is potential for agency in numerous autonomous, grassroots individuals and groups who can build "prefigurative" projects for non-violent radical social change. He defines "prefiguration" as "the potential for the given to contain the lineaments of what is to be", meaning that "a moment toward the future exists embedded in every point of the social organism where a need arises".

If "everything has prefigurative potential", Kovel notes that forms of potential ecological production will be "scattered", and thus suggests that "the task is to free them and connect them". While all "human ecosystems" have "ecosocialist potential", Kovel points out that ones such as the World Bank have low potential, whereas internally democratic anti-globalization "affinity groups" have a high potential through a dialectic that involves the "active bringing and holding together of negations", such as the group acting as an alternative institution ("production of an ecological/socialist alternative") and trying to shut down a G8 summit meeting ("resistance to capital"). Therefore, "practices that in the same motion enhance use-values and diminish exchange-values are the ideal" for eco-socialists.

Prefiguration

For Kovel, the main prefigurative steps "are that people ruthlessly criticize the capitalist system... and that they include in this a consistent attack on the widespread belief that there can be no alternative to it", which will then "deligitimate the system and release people into struggle". Kovel justifies this by stating that "radical criticism of the given... can be a material force", even without an alternative, "because it can seize the mind of the masses of people", leading to "dynamic" and "exponential", rather than "incremental" and "linear", victories that spread rapidly. Following this, he advocates the expansion of the dialectical eco-socialist potential of groups through sustaining the confrontation and internal cohesion of human ecosystems, leading to an "activation" of potentials in others that will "spread across the whole social field" as "a new set of orienting principles" that define an ideology or "'party-life' formation".

In the short-term, eco-socialists like Kovel advocate activities that have the "promise of breaking down the commodity form". This includes organizing labor, which is a "reconfiguring of the use-value of labor power"; forming cooperatives, allowing "a relatively free association of labor"; forming localised currencies, which he sees as "undercutting the value-basis of money"; and supporting "radical media" that, in his eyes, involve an "undoing of the fetishism of commodities". Arran Gare, Wall and Kovel have advocated economic localisation in the same vein as many in the Green movement, although they stress that it must be a prefigurative step rather than an end in itself.

Kovel also advises political parties attempting to "democratize the state" that there should be "dialogue but no compromise" with established political parties, and that there must be "a continual association of electoral work with movement work" to avoid "being sucked back into the system". Such parties, he believes, should focus on "the local rungs of the political system" first, before running national campaigns that "challenge the existing system by the elementary means of exposing its broken promises".

Kovel believes in building prefigurations around forms of production based on use values, which will provide a practical vision of a post-capitalist, post-statist system. Such projects include Indymedia ("a democratic rendering of the use-values of new technologies such as the Internet, and a continual involvement in wider struggle"), open-source software, Wikipedia, public libraries and many other initiatives, especially those developed within the anti-globalisation movement. These strategies, in Wall's words, "go beyond the market and the state" by rejecting the supposed dichotomy between private enterprise and state-owned production, while also rejecting any combination of the two through a mixed economy. He states that these present forms of "amphibious politics", which are "half in the dirty water of the present but seeking to move on to a new, unexplored territory".

Wall suggests that open source software, for example, opens up "a new form of commons regime in cyberspace", which he praises as production "for the pleasure of invention" that gives "access to resources without exchange". He believes that open source has "bypassed" both the market and the state, and could provide "developing countries with free access to vital computer software". Furthermore, he suggests that an "open source economy" means that "the barrier between user and provider is eroded", allowing for "cooperative creativity". He links this to Marxism and the notion of usufruct, asserting that "Marx would have been a Firefox user".

Internationalization of Prefiguration and the 'Eco-socialist Party'

Many eco-socialists have noted that the potential for building such projects is easier for media workers than for those in heavy industry because of the decline in trade unionism and the globalized division of labor which divides workers. However, Kovel believes that examples like the Christian Bruderhof Communities (despite elements of patriarchy that he attacks) show that "communistic" organizations can "survive rather well in a heavily industrialized market" if they are "protected" from the dependence on the market by "anti-capitalist intentionality". He further posits that class struggle is "internationalized in the face of globalization", as evidenced by a wave of strikes across the Global South in the first half of the year 2000; indeed, he says that "labor's most cherished values are already immanently ecocentric".

Kovel therefore thinks that these universalizing tendencies must lead to the formation of "a consciously 'Ecosocialist Party'" that is neither like a parliamentary or vanguardist party. Instead, Kovel advocates a form of political party "grounded in communities of resistance", where delegates from these communities form the core of the party's activists, and these delegates and the "open and transparent" assembly they form are subject to recall and regular rotation of members. He holds up the Zapatista Army of National Liberation (EZLN) and the Gaviotas movement as examples of such communities, which "are produced outside capitalist circuits" and show that "there can be no single way valid for all peoples".

Nonetheless, he also firmly believes in connecting these movements, stating that "ecosocialism will be international or it will be nothing" and hoping that the Ecosocialist Party can retain the autonomy of local communities while supporting them materially. With an ever-expanding party, Kovel hopes that "defections" by capitalists will occur, leading eventually to the armed forces and police who, in joining the revolution, will signify that "the turning point is reached".

'The Revolution' and Transition to Eco-socialism

'The Revolution' as envisaged by eco-socialists involves an immediate socio-political transition. Internationally, eco-socialists believe in a reform of the nature of money and the formation of a 'World People's Trade Organisation' (WPTO) that democratizes and improves world trade through the calculation of an 'Ecological Price' (EP) for goods. This would then be followed by a transformation of socioeconomic conditions towards ecological production, commons land and notions of usufruct (that seek to improve the common property possessed by society) to end private property. Eco-socialists assert that this must be carried out with adherence to non-violence.

The Immediate Aftermath of the Revolution

Eco-socialists like Kovel use the term "Eco-socialist revolution" to describe the transition to an eco-socialist world society. In the immediate socio-political transition, he believes that four groups will emerge from the revolution – revolutionaries, those "whose productive activity is directly compatible with ecological production" (such as nurses, schoolteachers, librarians, independent farmers and many other examples), those "whose pre-revolutionary practice was given over to capital" (including the bourgeoisie, advertising executives and more) and "the workers whose activity added surplus value to capitalist commodities".

In terms of political organisation, he advocates an "interim assembly" made up of the revolutionaries that can "devise incentives to make sure that vital functions are maintained" (such as short-term continuation of "differential remuneration" for labor), "handle the redistribution of social roles and assets", convene "in widespread locations", and send delegates to regional, state, national and international organisations, where every level has an "executive council" that is rotated and can be recalled. From there, he asserts that "productive communities" will "form the political as well as economic unit of society" and "organize others" to make a transition to eco-socialist production.

He adds that people will be allowed to be members of any community they choose with "associate membership" of others, such as a doctor having main membership of healthcare communities as a doctor and associate membership of child-rearing communities as a father. Each locality would, in Kovel's eyes, require one community that administered the areas of jurisdiction through an elected assembly. High-level assemblies would have additional "supervisory" roles over localities to monitor the development of ecosystemic integrity, and administer "society-wide services" like transport in "state-like functions", before the interim assembly can transfer responsibilities to "the level of the society as a whole through appropriate and democratically responsive committees".

Transnational Trade and Capital Reform

Part of the eco-socialist transition, in Kovel's eyes, is the reforming money to retain its use in "enabling exchanges" while reducing its functions as "a commodity in its own right" and "repository of value". He argues for directing money to "enhancement of use-values" through a "subsidization of use-values" that "preserves the functioning core of the economy while gaining time and space for rebuilding it". Internationally, he believes in the immediate cessation of speculation in currencies ("breaking down the function of money as commodity, and redirecting funds on use-values"), the cancellation of the debt of the Global South ("breaking the back of the value function" of money) and the redirecting the "vast reservoir of mainly phony value" to reparations and "ecologically sound development". He suggests the end of military aid and other forms of support to "comprador elites in the South" will eventually "lead to their collapse".

In terms of trade, Kovel advocates a 'World People's Trade Organization' (WPTO), "responsible to a confederation of popular bodies", in which "the degree of control over trade is... proportional to involvement with production", meaning that "farmers would have a special say over food trade" and so on. He posits that the WPTO should have an elected council that will oversee a reform of prices in favour of an 'Ecological Price' (EP) "determined by the difference between actual use-values and fully realized ones", thus having low tariffs for forms of ecological production like organic agriculture; he also envisages the high tariffs on non-ecological production providing subsidies to ecological production units.

The EP would also internalize the costs of current externalities (like pollution) and "would be set as a function of the distance traded", reducing the effects of long-distance transport like carbon emissions and increased packaging of goods. He thinks that this will provide a "standard of transformation" for non-ecological industries, like the automobile industry, thus spurring changes towards ecological production.

Ecological Production

Eco-socialists pursue "ecological production" that, according to Kovel, goes beyond the socialist vision of the emancipation of labor to "the realization of use-values and the appropriation of intrinsic value". He envisions a form of production in which "the making of a thing becomes part of the thing made" so that, using a high quality meal as an analogy, "pleasure would obtain for the cooking of the meal" - thus activities "reserved as hobbies under capitalism" would "compose the fabric of everyday life" under eco-socialism.

This, for Kovel, is achieved if labor is "freely chosen and developed with a fully realized use-value" achieved by a "negation" of exchange-value, and he exemplifies the Food Not Bombs project for adopting this. He believes that the notion of "mutual recognition for the process as well as the product" will avoid exploitation and hierarchy. With production allowing humanity to "live more directly and receptively embedded in nature", Kovel predicts that "a reorientation of human need" will occur that recognises ecological limits and sees technology as "fully participant in the life of eco-systems", thus removing it from profit-making exercises.

In the course on an Eco-socialist revolution, writers like Kovel advocate a "rapid conversion to ecosocialist production" for all enterprises, followed by "restoring ecosystemic integrity to the workplace" through steps like workers ownership. He then believes that the new enterprises can build "socially developed plans" of production for societal needs, such as efficient light-rail transport components. At the same time, Kovel argues for the transformation of essential but, under capitalism, non-productive labour, such as child care, into productive labour, "thereby giving reproductive labour a status equivalent to productive labour".

During such a transition, he believes that income should be guaranteed and that money will still be used under "new conditions of value... according to use and to the degree to which ecosystem integrity is developed and advanced by any particular production". Within this structure, Kovel asserts that markets and will become unnecessary – although "market phenomena" in personal exchanges and other small instances might be adopted – and communities and elected assemblies will democratically decide on the allocation of resources. Istvan Meszaros believes that such "genuinely planned and self-managed (as opposed to bureaucratically planned from above) productive activities" are essential if eco-socialism is to meet its "fundamental objectives".

Eco-socialists are quick to assert that their focus on "production" does not mean that there will be an increase in production and labor under Eco-socialism. Kovel thinks that the emancipation of labor and the realization of use-value will allow "the spheres of work and culture to be reintegrated". He cites the example of Paraguayan Indian communities (organised by Jesuits) in the eighteenth century who made sure that all community members learned musical instruments, and had labourers take musical instruments to the fields and takes turns playing music or harvesting.

Commons, Property and 'Usufruct'

Most eco-socialists, including Guha and Martinez-Alier, echo subsistence eco-feminists like Vandana Shiva when they argue for the restoration of commons land over private property. They blame ecological degradation on the inclination to short-term, profit-inspired decisions inherent within a market system. For them, privatization of land strips people of their local communal resources in the name of creating markets for neo-liberal globalisation, which benefits a minority. In their view, successful commons systems have been set up around the world throughout history to manage areas cooperatively, based on long-term needs and sustainability instead of short-term profit.

Many eco-socialists focus on a modified version of the notion of 'Usufruct' to replace capitalist private property arrangements. As a legal term, Usufruct refers to the legal right to use and derive profit or benefit from property that belongs to another person, as long as the property is not damaged. According to eco-socialists like Kovel, a modern interpretation of the idea is "where one uses, enjoys – and through that, improves – another's property", as its Latin etymology "condenses the two meanings of use – as in use-value, and enjoyment – and as in the gratification expressed in freely associated labour". The idea, according to Kovel, has roots in the Code of Hammurabi and was first mentioned in Roman law "where it applied to ambiguities between masters and slaves with respect to property"; it also features in Islamic Sharia law, Aztec law and the Napoleonic Code.

Crucially for eco-socialists, Marx mentioned the idea when he stated that human beings are no more than the planet's "usufructaries, and, like *boni patres familias*, they must hand it down to succeeding generations in an improved condition". Kovel and others have taken on this reading, asserting that, in an eco-socialist society, "everyone will have... rights of use and ownership over those means of production necessary to express the creativity of human nature", namely "a place of one's own" to decorate to personal taste, some personal possessions, the body and its attendant sexual and reproductive rights.

However, Kovel sees property as "self-contradictory" because individuals emerge "in a tissue of social relations" and "nested circles", with the self at the centre and extended circles where "issues of sharing arise from early childhood on". He believes that "the full self is enhanced more by giving than by taking" and that eco-socialism is realized when material possessions weigh "lightly" upon the self – thus restoration of use-value allows things to be taken "concretely and sensuously" but "lightly, since things are enjoyed for themselves and not as buttresses for a shaky ego".

This, for Kovel, reverses what Marxists see as the commodity fetishism and atomization of individuals (through the "unappeasable craving" for "having and excluding others from having") under capitalism. Under eco-socialism, he therefore believes that enhancement of use-value will lead to differentiated ownership between the individual and the collective, where there are "distinct limits on the amount of property individuals control" and no-one can take control of resources that "would permit the alienation of means of production from another". He then hopes that the "hubris" of the notion of "ownership of the planet" will be replaced with usufruct.

Non-violence

Most eco-socialists are involved in peace and antiwar movements, and eco-socialist writers, like Kovel, generally believe that "violence is the rupturing of ecosystems" and is therefore "deeply contrary to ecosocialist values". Kovel believes that revolutionary movements must prepare for post-revolutionary violence from counter-revolutionary sources by "prior development of the democratic sphere" within the movement, because "to the degree that people are capable of self-government, so will they turn away from violence and retribution" for "a self-governed people cannot be pushed around by any alien government". It is therefore essential, in Kovel's view, that the revolution "takes place in" or spreads quickly to the United States, which "is capital's gendarme and will crush any serious threat", and that revolutionaries reject the death penalty and retribution against former opponents or counter-revolutionaries.

Criticisms of Eco-socialism

While in many ways the criticisms of eco-socialism combine the traditional criticisms of both socialism and Green politics, there are unique critiques of eco-socialism, which are largely from within the traditional Socialist or Green movements themselves, along with conservative criticisms.

Some socialists are critical of the word 'eco-socialism'. David Reilly, who questions whether his argument is improved by the use of an "exotic word", argues instead that the "real socialism" is "also a green or 'eco'" one that you get to "by dint of struggle". Other socialists, like Paul Hampton of the Alliance for Workers' Liberty (a British third camp socialist party), eco-socialism as "classless ecology", wherein eco-socialists have "given up on the working class" as the privileged agent of struggle by "borrowing bits from Marx but missing the locus of Marxist politics".

Writing in *Capitalism Nature Socialism*, Doug Boucher, Peter Caplan, David Schwartzman and Jane Zara criticise eco-socialists in general (and Joel Kovel in particular) for a deterministic "catastrophism" that overlooks "the countervailing tendencies of both popular struggles and the efforts of capitalist governments to rationalize the system" and the "accomplishments of the labor movement" that "demonstrate that despite the interests and desires of capitalists, progress toward social justice is possible". They argue that an ecological socialism must be "built on hope, not fear".

Conservatives have criticised the perceived opportunism of left-wing groups who have increased their focus on green issues since the fall of Communism. Fred L. Smith Jr., President of the Competitive Enterprise Institute think-tank, exemplifies the conservative critique of left Greens, attacking the "pantheism" of the Green movement and conflating "eco-paganism" with eco-socialism. Like many conservative critics, Smith uses the term 'eco-socialism' to attack non-socialist environmentalists for advocating restrictions on the free market, although he does recognise and condemn the influence of socialist ideals on many in the Green movement who reject private property, and instead advocates market-based solutions to ecological problems. He nevertheless wrongly claims that eco-socialists endorse "the Malthusian view of the relationship between man and nature", and states that Al Gore, a former Democratic Party Vice President of the United States and now a climate change campaigner, is an eco-socialist, despite the fact that Gore has never used this term and is not recognised as a such by other followers of either Green politics or socialism.

Some environmentalists and conservationists have criticised eco-socialism from within the Green movement. In a review of Joel Kovel's *The Enemy of Nature*, David M. Johns criticises eco-socialism for not offering "suggestions about near term conservation policy" and focusing exclusively on long-term societal transformation. Johns believes that species extinction "started much earlier" than capitalism and suggests that eco-socialism neglects the fact that an ecological society will need to transcend the destructiveness found in "all large-scale societies". the very tendency that Kovel himself attacks among capitalists and traditional leftists who attempt to reduce nature to "linear" human models. Johns questions whether non-hierarchical social systems can provide for billions of people, and criticises eco-socialists for neglecting issues of population pressure. Furthermore, Johns describes Kovel's argument that human hierarchy is founded on raiding to steal women as "archaic".

Green Anarchism

Green anarchism (or eco-anarchism) is a school of thought within anarchism which puts a particular emphasis on environmental issues. A green anarchist theory is normally one that extends anarchist ideology beyond a critique of human interactions, and includes a critique of the interactions between humans and non-humans as well. This often culminates in an anarchist revolutionary praxis that is not merely dedicated to human liberation, but also to some form of nonhuman liberation, and that aims to bring about an environmentally sustainable anarchist society.

Important early influences were Henry David Thoreau, Leo Tolstoy and Élisée Reclus. In the late 19th century there emerged anarcho-naturism as the fusion of anarchism and naturist philosophies within individualist anarchist circles in France, Spain, Cuba, and Portugal. Important contemporary currents (some of which may be mutually exclusive) include anarcho-primitivism, which offers a critique of technology and argues that anarchism is best suited to pre-"civilised" ways of life; veganarchism, which argues that human liberation and animal liberation are inseparable; and social ecology, which argues that the hierarchical domination of nature by human stems from the hierarchical domination of human by human.

Early Ecoanarchism

Henry David Thoreau

Henry David Thoreau, influential early green-anarchist who wrote *Walden*

Anarchism started to have an ecological view mainly in the writings of American anarchist and transcendentalist Henry David Thoreau. In his book *Walden* he advocates simple living and self-sufficiency among natural surroundings in resistance to the advancement of industrial civilization. The work is part personal declaration of independence, social experiment, voyage of spiritual discovery, satire, and manual for self-reliance. First published in 1854, it details Thoreau's experiences over the course of two years, two months, and two days in a cabin he built near Walden Pond, amidst woodland owned by his friend and mentor Ralph Waldo Emerson, near Concord, Massachusetts. The book compresses the time into a single calendar year and uses passages of four seasons to symbolize human development. By immersing himself in nature, Thoreau hoped to gain a more objective understanding of society through personal introspection. Simple living and self-sufficiency were Thoreau's other goals, and the whole project was inspired by transcendentalist philosophy, a central theme of the American Romantic Period. As Thoreau made clear in his book, his cabin was not in wilderness but at the edge of town, about two miles (3 km) from his family home.

As such "Many have seen in Thoreau one of the precursors of ecologism and anarcho-primitivism represented today in John Zerzan. For George Woodcock this attitude can be also motivated by certain idea of resistance to progress and of rejection of the growing materialism which is the nature of American society in the mid 19th century." John Zerzan himself included the text "Excursions" (1863) by Thoreau in his edited compilation of writings called *Against civilization: Readings and reflections* from 1999.

Élisée Reclus

Élisée Reclus (15 March 1830 – 4 July 1905), also known as Jacques Élisée Reclus, was a renowned

French geographer, writer and anarchist. He produced his 19-volume masterwork *La Nouvelle Géographie universelle, la terre et les hommes* ("Universal Geography"), over a period of nearly 20 years (1875–1894). In 1892 he was awarded the prestigious Gold Medal of the Paris Geographical Society for this work, despite his having been banished from France because of his political activism.

Élisée Reclus, French anarchist geographer and early environmentalist

According to Kirkpatrick Sale:

His geographical work, thoroughly researched and unflinchingly scientific, laid out a picture of human-nature interaction that we today would call bioregionalism. It showed, with more detail than anyone but a dedicated geographer could possibly absorb, how the ecology of a place determined the kinds of lives and livelihoods its denizens would have and thus how people could properly live in self-regarding and self-determined bioregions without the interference of large and centralized governments that always try to homogenize diverse geographical areas.

For the authors of An Anarchist FAQ Reclus "argued that a "secret harmony exists between the earth and the people whom it nourishes, and when imprudent societies let themselves violate this harmony, they always end up regretting it." Similarly, no contemporary ecologist would disagree with his comments that the "truly civilised man [and women] understands that his [or her] nature is bound up with the interest of all and with that of nature. He [or she] repairs the damage caused by his predecessors and works to improve his domain."

Reclus advocated nature conservation and opposed meat-eating and cruelty to animals. He was a vegetarian. As a result, his ideas are seen by some historians as anticipating the modern social ecology and animal rights movements. Shortly before his death, Reclus completed *L'Homme et la terre* (1905). In it, he added to his previous greater works by considering humanity's development relative to its geographical environment. Reclus was also an early proponent of naturism.

Anarcho-naturism

In the late 19th century Anarchist naturism appeared as the union of anarchist and naturist philosophies. Mainly it had importance within individualist anarchist circles in Spain, France, Portugal, and Cuba.

Anarcho-naturism advocated vegetarianism, free love, nudism and an ecological world view within anarchist groups and outside them. Anarcho-naturism promoted an ecological worldview, small ecovillages, and most prominently nudism as a way to avoid the artificiality of the industrial mass society of modernity. Naturist individualist anarchists saw the individual in his biological, physical and psychological aspects and tried to eliminate social determinations. Important promoters of this were Henri Zisly and Emile Gravelle who collaborated in *La Nouvelle Humanité* followed by *Le Naturien, Le Sauvage, L'Ordre Naturel, & La Vie Naturelle.*

France

Richard D. Sonn comments on the influence of naturist views in the wider French anarchist movement:

> In her memoir of her anarchist years that was serialized in *Le Matin* in 1913, Rirette Maîtrejean made much of the strange food regimens of some of the *compagnons*. She described the "tragic bandits" of the Bonnot gang as refusing to eat meat or drink wine, preferring plain water. Her humorous comments reflected the practices of the "naturist" wing of individualist anarchists who favored a simpler, more "natural" lifestyle centered on a vegetarian diet. In the 1920s, this wing was expressed by the journal *Le Néo-Naturien, Revue des Idées Philosophiques et Naturiennes.* Contributors condemned the fashion of smoking cigarettes, especially by young women; a long article of 1927 actually connected cigarette smoking with cancer! Others distinguished between vegetarians, who foreswore the eating of meat, from the stricter "vegetalians," who ate nothing but vegetables. An anarchist named G. Butaud, who made this distinction, opened a restaurant called the Foyer Végétalien in the nineteenth arrondissement in 1923. Other issues of the journal included vegetarian recipes. In 1925, when the young anarchist and future detective novelist Léo Malet arrived in Paris from Montpellier, he initially lodged with anarchists who operated another vegetarian restaurant that served only vegetables, with neither fish nor eggs. Nutritional concerns coincided with other means of encouraging health bodies, such as nudism and gymnastics. For a while in the 1920s, after they were released from jail for antiwar and birth-control activities, Jeanne and Eugène Humbert retreated to the relative safety of the "integral living" movement that promoted nude sunbathing and physical fitness, which were seen as integral aspects of health in the Greek sense of *gymnos*, meaning nude. This back-to-nature, primitivist current was not a monopoly of the left; the same interests were echoed by right-wing Germans in the interwar era. In France, however, these proclivities were mostly associated with anarchists, insofar as they suggested an ideal of self-control and the rejection of social taboos and prejudices.

Henri Zisly

Henri Zisly (born in Paris, November 2, 1872; died in 1945) was a French individualist anarchist and naturist. He participated alongside Henri Beylie and Émile Gravelle in many journals such as *La Nouvelle Humanité* and *La Vie Naturelle*, which promoted anarchist-naturism. In 1902, he was one of the main initiators, alongside Georges Butaud and Sophie Zaïkowska, of the cooperative *Colonie de Vaux* established in Essômes-sur-Marne, in l'Aisne.

Zisly's political activity, "primarily aimed at supporting a return to 'natural life' through writing and practical involvement, stimulated lively confrontations within and outside the anarchist environment. Zisly vividly criticized progress and civilization, which he regarded as 'absurd, ignoble, and filthy.' He openly opposed industrialization, arguing that machines were inherently authoritarian, defended nudism, advocated a non-dogmatic and non-religious adherence to the 'laws of nature,' recommended a lifestyle based on limited needs and self-sufficiency, and disagreed with vegetarianism, which he considered 'anti-scientific.'"

Cuba

The historian Kirwin R. Schaffer in his study of Cuban anarchism reports anarcho-naturism as "A third strand within the island's anarchist movement" alongside anarcho-communism and anarcho-syndicalism. Naturism was a global alternative health and lifestyle movement. Naturists focused on redefining one's life to live simply, eat cheap but nutritious vegetarian diets, and raise one's own food if possible. The countryside was posited as a romantic alternative to urban living, and some naturists even promoted what they saw as the healthful benefits of nudism. Globally, the naturist movement counted anarchists, liberals, and socialists as its followers. However, in Cuba a particular "anarchist" dimension evolved led by people like Adrián del Valle, who spearheaded the Cuban effort to shift naturism's focus away from only individual health to naturism having a "social emancipatory" function."

Schaffer reports the influence that anarcho-naturism had outside naturists circles. So "For instance, nothing inherently prevented an anarcho-syndicalist in the Havana restaurant workers' union from supporting the alternative health care programs of the anarcho-naturists and seeing those alternative practices as "revolutionary."". "Anarcho-naturists promoted a rural ideal, simple living, and being in harmony with Nature as ways to save the laborers from the increasingly industrialized character of Cuba. Besides promoting an early twentieth-century "back-to-the-land" movement, they used these romantic images of Nature to illustrate how far removed a capitalist industrialized Cuba had departed from an anarchist view of natural harmony." The main propagandizer in Cuba of anarcho-naturism was the Catalonia born "Adrián del Valle (aka Palmiro de Lidia)...Over the following decades, Del Valle became a constant presence in not only the anarchist press that proliferated in Cuba but also mainstream literary publications...From 1912 to 1913 he edited the freethinking journal *El Audaz*. Then he began his largest publishing job by helping to found and edit the monthly alternative health magazine that followed the anarcho-naturist line *Pro-Vida*.

Spain

Isaac Puente, spanish anarchist naturist and anarcho-communist

Anarcho-naturism was quite important at the end of the 1920s in the spanish anarchist movement In France, later important propagandists of anarcho-naturism include Henri Zisly and Émile Gravelle whose ideas were important in individualist anarchist circles in Spain, where Federico Urales (pseudonym of Joan Montseny) promoted the ideas of Gravelle and Zisly in *La Revista Blanca* (1898–1905).

> The linking role played by the Sol y Vida group was very important. The goal of this group was to take trips and enjoy the open air. The Naturist athenaeum, Ecléctico, in Barcelona, was the base from which the activities of the group were launched. First Etica and then Iniciales, which began in 1929, were the publications of the group, which lasted until the Spanish Civil War. We must be aware that the naturist ideas expressed in them matched the desires that the libertarian youth had of breaking up with the conventions of the bourgeoisie of the time. That is what a young worker explained in a letter to Iniciales. He writes it under the odd pseudonym of silvestre del campo (wild man in the country). "I find great pleasure in being naked in the woods, bathed in light and air, two natural elements we cannot do without. By shunning the humble garment of an exploited person, (garments which, in my opinion, are the result of all the laws devised to make our lives bitter), we feel there no others left but just the natural laws. Clothes mean slavery for some and tyranny for others. Only the naked man who rebels against all norms, stands for anarchism, devoid of the prejudices of outfit imposed by our money-oriented society."

The "relation between Anarchism and Naturism gives way to the Naturist Federation, in July 1928, and to the lV Spanish Naturist Congress, in September 1929, both supported by the Libertarian Movement. However, in the short term, the Naturist and Libertarian movements grew apart in their conceptions of everyday life. The Naturist movement felt closer to the Libertarian individualism of some French theoreticians such as Henri Ner (real name of Han Ryner) than to the revolutionary goals proposed by some Anarchist organisations such as the FAI, (Federación Anarquista Ibérica)". This ecological tendency in Spanish anarchism was strong enough as to call the attention of the CNT–FAI in Spain. Daniel Guérin in *Anarchism: From Theory to Practice* reports:

> Spanish anarcho-syndicalism had long been concerned to safeguard the autonomy of what it called "affinity groups." There were many adepts of naturism and vegetarianism among its members, especially among the poor peasants of the south. Both these ways of living were considered suitable for the transformation of the human being in preparation for a libertarian society. At the Saragossa congress the members did not forget to consider the fate of groups of naturists and nudists, "unsuited to industrialization." As these groups would be unable to supply all their own needs, the congress anticipated that their delegates to the meetings of the confederation of communes would be able to negotiate special economic agreements with the other agricultural and industrial communes. On the eve of a vast, bloody, social transformation, the CNT did not think it foolish to try to meet the infinitely varied aspirations of individual human beings.

Isaac Puente

Isaac Puente was an influential Spanish anarchist during the 1920s and 1930s and an important propagandist of anarcho-naturism, was a militant of both the CNT anarcho-syndicalist trade union and Iberian Anarchist Federation. He published the book *El Comunismo Libertario y otras proclamas insurreccionales y naturistas* (en:*Libertarian Communism and other insurrectionary and naturist proclamations*) in 1933, which sold around 100,000 copies, and wrote the final document for the Extraordinary Confederal Congress of Zaragoza of 1936 which established the main political line for the CNT for that year. Puente was a doctor who approached his medical practice from a naturist point of view. He saw naturism as an integral solution for the working classes, alongside Neo-Malthusianism, and believed it concerned the living being while anarchism addressed the social being. He believed capitalist societies endangered the well-being of humans from both a socioeconomic and sanitary viewpoint, and promoted anarcho-communism alongside naturism as a solution.

Other Countries

Naturism also met anarchism in the United Kingdom. "In many of the alternative communities established in Britain in the early 1900s nudism, anarchism, vegetarianism and free love were accepted as part of a politically radical way of life. In the 1920s the inhabitants of the anarchist community at Whiteway, near Stroud in Gloucestershire, shocked the conservative residents of the area with their shameless nudity." In Italy, during the IX Congress of the Italian Anarchist Federation in Carrara in 1965, a group decided to split off from this organization and created the *Gruppi di Iniziativa Anarchica*. In the seventies, it was mostly composed of "veteran individualist anarchists with an orientation of pacifism, naturism, etc". American anarcho-syndicalist Sam Dolgoff shows some of the criticism that some people on the other anarchist currents at the time had for anarcho-naturist tendencies. "Speaking of life at the Stelton Colony of New York in the 1930s, noted with disdain that it, "like other colonies, was infested by vegetarians, naturists, nudists, and other cultists, who sidetracked true anarchist goals." One resident "always went barefoot, ate raw food, mostly nuts and raisins, and refused to use a tractor, being opposed to machinery, and he didn't want to abuse horses, so he dug the earth himself." Such self-proclaimed anarchists were in reality "ox-cart anarchists," Dolgoff said, "who opposed organization and wanted to return to a simpler life." In an interview with Paul Avrich before his death, Dolgoff also grumbled, "I am sick and tired of these half-assed artists and poets who object to organization and want only to play with their belly buttons."

Leo Tolstoy and Tolstoyanism

Tolstoy dressed in peasant clothing, by Ilya Repin (1901)

Russian christian anarchist and anarcho-pacifist Leo Tolstoy is also recognized as an early influence in green anarchism. The novelist was struck by the description of Christian, Buddhist, and Hindu ascetic renunciation as being the path to holiness. After reading passages such as the following, which abound in Schopenhauer's ethical chapters, the Russian nobleman chose poverty and formal denial of the will:

*But this very necessity of involuntary suffering (by poor people) for eternal salvation is also expressed by that utterance of the Savior (Matthew 19:24): "It is easier for a camel to go through the eye of a needle, than for a rich man to enter into the kingdom of God." Therefore those who were greatly in earnest about their eternal salvation, chose voluntary poverty when fate had denied this to them and they had been born in wealth. Thus Buddha Sakyamuni was born a prince, but voluntarily took to the mendicant's staff; and Francis of Assisi, the founder of the mendicant orders who, as a youngster at a ball, where the daughters of all the notabilities were sitting together, was asked: "Now Francis, will you not soon make your choice from these beauties?" and who replied: "I have made a far more beautiful choice!" "Whom?" "*La povertà *(poverty)": whereupon he abandoned every thing shortly afterwards and wandered through the land as a mendicant.*

Despite his misgivings about anarchist violence, Tolstoy took risks to circulate the prohibited publications of anarchist thinkers in Russia, and corrected the proofs of Kropotkin's "Words of a Rebel", illegally published in St Petersburg in 1906. Tolstoy was enthused by the economic thinking of Henry George, incorporating it approvingly into later works such as *Resurrection*, the book that played a major factor in his excommunication. Tolstoyans identify themselves as Christians, but do not generally belong to an institutional Church. They attempt to live an ascetic and simple life, preferring to be vegetarian, non-smoking, teetotal and chaste. Tolstoyans are considered Christian pacifists and advocate nonresistance in all circumstances. They do not support or participate in the government which they consider immoral, violent and corrupt. Tolstoy rejected the state (as it only exists on the basis of physical force) and all institutions that are derived from it - the police, law courts and army. Tolstoy influenced Mohandas Karamchand Gandhi who set up a cooperative colony called Tolstoy Farm near Johannesburg, South Africa, having been inspired by Tolstoy's ideas. The colony comprising 1,100 acres (4.5 km^2) was funded by the Gandhian Herman Kallenbach and placed at the disposal of the satyagrahis from 1910. He also inspired similar communal experiments in the United States where the residents were also influenced by the views of Henry George and Edward Bellamy., as well as in Russia, England and the Netherlands.

Mid twentieth Century

Several anarchists from the mid twentieth century, including Herbert Read, Ethel Mannin, Leopold Kohr, and Paul Goodman, also held proto-environmental views linked to their anarchism. Mannin's 1944 book *Bread and Roses: A Utopian Survey and Blue-Print* has been described by anarchist historian Robert Graham as setting forth "an ecological vision in opposition to the prevailing and destructive industrial organization of society".

Leopold Kohr

Leopold Kohr (5 October 1909 in Oberndorf bei Salzburg, Austria – 26 February 1994 in Gloucester, England) was an economist, philosopher and political scientist known both for his opposition to the "cult of bigness" in social organization and as one of those who initiated the *small is beautiful* movement. For almost twenty years he was Professor of Economics and Public Administration at the University of Puerto Rico. He described himself as a "philosophical anarchist." In 1937, Kohr became a freelance correspondent during the Spanish Civil War, where he was impressed by the limited, self-contained governments of the separatist states of Catalonia and Aragon, as well as the small Spanish anarchist city states of Alcoy and Caspe. In his first published essay "Disunion Now: A Plea for a Society based upon Small Autonomous Units", published in Commonweal in 1941, Kohr wrote about a Europe at war: "We have ridiculed the many little states, now we are terrorized by their few successors." He called for the breakup of Europe into hundreds of city states. Kohr developed his ideas in a series of books, including *The Breakdown of Nations* (1957), *Development without Aid* (1973) and *The Overdeveloped Nations* (1977). From Leopold Kohr's most popular work *The Breakdown of Nations*:

... there seems to be only one cause behind all forms of social misery: bigness. Oversimplified as this may seem, we shall find the idea more easily acceptable if we consider that bigness, or oversize, is really much more than just a social problem. It appears to be the one and only problem permeating all creation. Whenever something is wrong, something is too big. ... And if the body of a people becomes diseased with the fever of aggression, brutality, collectivism, or massive idiocy, it is not because it has fallen victim to bad leadership or mental derangement. It is because human beings, so charming as individuals or in small aggregations, have been welded into overconcentrated social units.

Later in his academic and writing career he protested the "cult of bigness" and economic growth and promoted the concept of human scale and small community life. He argued that massive external aid to poorer nations stifled local initiatives and participation. His vision called for a dissolution of centralized political and economic structures in favor of local control. Kohr was an important inspiration to the Green, bioregional, Fourth World, decentralist, and anarchist movements, Kohr contributed often to John Papworth's `Journal for the Fourth World', Resurgence. One of Kohr's students was economist E. F. Schumacher, another prominent influence on these movements, whose best selling book *Small Is Beautiful* took its title from one of Kohr's core principles. Similarly, his ideas inspired Kirkpatrick Sale's books *Human Scale* (1980) and *Dwellers in the Land: The Bioregional Vision* (1985). Sale arranged the first American publication of *The Breakdown of Nations* in 1978 and wrote the foreword.

Murray Bookchin

Murray Bookchin (January 14, 1921 – July 30, 2006) was an American libertarian socialist author, orator, and philosopher. In 1958, Murray Bookchin defined himself as an anarchist, seeing parallels between anarchism and ecology. His first book, *Our Synthetic Environment,* was published under the pseudonym Lewis Herber in 1962, a few months before Rachel Carson's *Silent Spring.* The book described a broad range of environmental ills but received little attention because of its political radicalism. His groundbreaking essay "Ecology and Revolutionary Thought" introduced ecology as a concept in radical politics. In 1968 he founded another group that published the influential *Anarchos* magazine, which published that and other innovative essays on post-scarcity and on ecological technologies such as solar and wind energy, and on decentralization and miniaturization. Lecturing throughout the United States, he helped popularize the concept of ecology to the counterculture.

Post-Scarcity Anarchism is a collection of essays written by Murray Bookchin and first published in 1971 by Ramparts Press. It outlines the possible form anarchism might take under conditions of post-scarcity. It is one of Bookchin's major works, and its radical thesis provoked controversy for being utopian and messianic in its faith in the liberatory potential of technology. Bookchin argues that post-industrial societies are also post-scarcity societies, and can thus imagine "the fulfillment of the social and cultural potentialities latent in a technology of abundance". The self-administration of society is now made possible by technological advancement and, when technology is used in an ecologically sensitive manner, the revolutionary potential of society will be much changed. In 1982, his book *The Ecology of Freedom* had a profound impact on the emerging ecology movement, both in the United States and abroad. He was a principal figure in the Burlington Greens in 1986-90, an ecology group that ran candidates for city council on a program to create neighborhood democracy. In *From Urbanization to Cities* (originally published in 1987 as *The Rise of Urbanization and the Decline of Citizenship*), Bookchin traced the democratic traditions that influenced his political philosophy and defined the implementation of the libertarian municipalism concept. A few years later *The Politics of Social Ecology*, written by his partner of 20 years, Janet Biehl, briefly summarized these ideas.

Jacques Ellul

Jacques Ellul (January 6, 1912 – May 19, 1994) was a French philosopher, law professor, sociologist, lay theologian, and Christian anarchist. He wrote several books about Christianity, the technological society, propaganda, and the interaction between religion and politics. Professor of History and the Sociology of Institutions on the Faculty of Law and Economic Sciences at the University of Bordeaux, he authored 58 books and more than a thousand articles over his lifetime in all, the dominant theme of which has been the threat to human freedom and religion created by modern technique. The Ellulian concept of technique is briefly defined within the "Notes to Reader" section of *The Technological Society* (1964). What many consider to be Ellul's most important work, *The Technological Society* (1964) was originally titled: *La Technique: L'enjeu du siècle* (literally, "The Stake of the Century"). In it, Ellul set forth seven characteristics of modern technology that make efficiency a necessity: rationality, artificiality, automatism of technical choice, self-augmentation, monism, universalism, and autonomy.

For Ellul the rationality of technique enforces logical and mechanical organization through division of labor, the setting of production standards, etc. And it creates an artificial system which "eliminates or subordinates the natural world." Today, he argues, the technological society is generally held sacred (cf. Saint Steve Jobs). Since he defines technique as "the totality of methods rationally arrived at, and having absolute efficiency (for a given stage of development) in every field of human activity", it is clear that his sociological analysis focuses not on the society of machines as such, but on the society of "efficient techniques".

Contemporary Developments

Notable contemporary writers espousing green anarchism include Layla AbdelRahim, Derrick Jensen, Jaime Semprun, George Draffan, John Zerzan, Starhawk and Alan Carter.

Social Ecology and Communalism

Social ecology is closely related to the work and ideas of Murray Bookchin and influenced by anarchist Peter Kropotkin. Social ecologists assert that the present ecological crisis has its roots in human social problems, and that the domination of human-over-nature stems from the domination of human-over-human.

Bookchin later developed a political philosophy to complement social ecology which he called "Communalism" (spelled with a capital "C" to differentiate it from other forms of communalism). While originally conceived as a form of Social anarchism, he later developed Communalism into a separate ideology which incorporates what he saw as the most beneficial elements of Anarchism, Marxism, syndicalism, and radical ecology.

Politically, Communalists advocate a network of directly democratic citizens' assemblies in individual communities/cities organized in a confederal fashion. This method used to achieve this is called Libertarian Municipalism which involves the establishment of face-to-face democratic institutions which are to grow and expand confederally with the goal of eventually replacing the nation-state.

Janet Biehl (born 1953) is a writer associated with social ecology, the body of ideas developed and publicized by Murray Bookchin. In 1986, she attended the Institute for Social Ecology and there, began a collaborative relationship with Bookchin, working intensively with him over the next two decades in the explication of social ecology from their shared home in Burlington, Vermont.

From 1987 to 2000, she and Bookchin co-wrote and co-published the theoretical newsletter *Green Perspectives*, later renamed *Left Green Perspectives*. She is the editor and compiler of *The Murray Bookchin Reader* (1997); the author of *The Politics of Social Ecology: Libertarian Municipalism* (1998) and *Rethinking Ecofeminist Politics* (1991); and coauthor (with Peter Staudenmaier) of *Ecofascism: Lessons from the German Experience* (1995).

Green Anarchist

The magazine *Green Anarchist* was for a while the principal voice in the UK advocating green anarchism, an explicit fusion of libertarian socialist and ecological thinking. Founded after the 1984 Stop the City protests, the magazine was launched in the summer of that year by an editorial collective consisting of Alan Albon, Richard Hunt and Marcus Christo. Early issues featured a range of broadly anarchist and ecological ideas, bringing together groups and individuals as varied as Class War, veteran anarchist writer Colin Ward, anarcho-punk band Crass, as well as the Peace Convoy, anti-nuclear campaigners, animal rights activists and so on. However the diversity that many saw as the publication's greatest strength quickly led to irreconcilable arguments between the essentially pacifist approach of Albon and Christo, and the advocacy of violent confrontation with the State favoured by Hunt. During the 1990s *Green Anarchist* came under the helm of an editorial collective that included Paul Rogers, Steve Booth and others, during which period the publication became increasingly aligned with primitivism, an anti-civilization philosophy advocated by writers such as John Zerzan and Fredy Perlman. Starting in 1995, Hampshire Police began a series of at least 56 raids, code named 'Operation Washington', that eventually resulted in the August to November 1997 Portsmouth trial of Green Anarchist editors Booth, Saxon Wood, Noel Molland and Paul Rogers, as well as Animal Liberation Front (ALF) Press Officer Robin Webb and Animal Liberation Front Supporters Group (ALFSG) newsletter editor Simon Russell. The defendants organised the GANDALF

Defence campaign. Three of the editors of Green Anarchist, Noel Molland, Saxon Wood and Booth were jailed for 'conspiracy to incite'. However, all three were shortly afterwards released on appeal.

Cover of the first issue of 'Green Anarchist' magazine (Summer 1984), featuring artwork by then editor Richard Hunt

Fredy Perlman

Fredy Perlman (August 20, 1934 – July 26, 1985) was a Czech-born, naturalised American author, publisher and militant. His most popular work, the book *Against His-Story, Against Leviathan!*, details the rise of state domination with a retelling of history through the Hobbesian metaphor of the Leviathan. The book remains a major source of inspiration for anti-civilization perspectives in contemporary anarchism, most notably on the thought of philosopher John Zerzan.

Anarcho-primitivism

John Zerzan, anarcho-primitivism theorist

Anarcho-primitivism is an anarchist critique of the origins and progress of civilization. According to anarcho-primitivism, the shift from hunter-gatherer to agricultural subsistence gave rise to social stratification, coercion, and alienation. Anarcho-primitivists advocate a return to non-"civilized" ways of life through deindustrialisation, abolition of the division of labour or specialization, and abandonment of large-scale organization technologies. There are other non-anarchist forms of primitivism, and not all primitivists point to the same phenomenon as the source of modern, civilized problems. Anarcho-primitivists are often distinguished by their focus on the praxis of achieving a feral state of being through "rewilding".

John Zerzan

John Zerzan is an American anarchist and primitivist philosopher and author. His works criticize agricultural civilization as inherently oppressive, and advocate drawing upon the ways of life of hunter gatherers as an inspiration for what a free society should look like. Some subjects of his criticism include domestication, language, symbolic thought (such as mathematics and art) and the concept of time.

His five major books are *Elements of Refusal* (1988), *Future Primitive and Other Essays* (1994), *Running on Emptiness* (2002), *Against Civilization: Readings and Reflections* (2005) and *Twilight of the Machines* (2008). Zerzan was one of the editors of *Green Anarchy*, a controversial journal of anarcho-primitivist and insurrectionary anarchist thought. He is also the host of *Anarchy Radio* in Eugene on the University of Oregon's radio station KWVA. He has also served as a contributing editor at *Anarchy Magazine* and has been published in magazines such as *AdBusters*. He does extensive speaking tours around the world, and is married to an independent consultant to museums and other nonprofit organizations. In 1974, Black and Red Press published *Unions Against Revolution* by Spanish ultra-left theorist Grandizo Munis that included an essay by Zerzan which previously appeared in the journal *Telos*. Over the next 20 years, Zerzan became intimately involved with the *Fifth Estate*, *Anarchy: A Journal of Desire Armed*, *Demolition Derby* and other anarchist periodicals. He began to question civilization in the early 80's, after having sought to confront issues around the neutrality of technology and division of labour, at the time when Fredy Perlman was making similar conclusions.

Green Anarchy

Green Anarchy was a magazine published by a collective located in Eugene, Oregon. The magazine's focus was primitivism, post-left anarchy, radical environmentalism, African American struggles, anarchist resistance, indigenous resistance, earth and animal liberation, anti-capitalism and supporting political prisoners. It had a circulation of 8,000, partly in prisons, the prison subscribers given free copies of each issue as stated in the magazine. Green Anarchy was started in 2000 and in 2009 the Green Anarchy website shut down, leaving a final, brief message about the cessation of the magazine's publication. The subtitle of the magazine is "An Anti-Civilization Journal of Theory and Action". Author John Zerzan was one of the publication's editors.

Species Traitor

Species Traitor is a sporadically published journal of insurrectionary anarcho-primitivism. It is printed as a project of Black and Green Network and edited by anarcho-primitivist writer, Kevin

Tucker. ST was initially labeled as a project of the Coalition Against Civilization (CAC) and the Black and Green Network (BAG). The CAC was started towards the end of 1999 in the aftermath of the massive street protests in Eugene (Reclaim the Streets) and in Seattle (WTO) of that year. That aftermath gave a new voice and standing for green anarchist and anarcho-primitivist writers and viewpoints within both the anarchist milieu and the culture at large. The first issue came out in winter of 2000-2001 (currently out of print) and contained a mix of reprints and some original articles from Derrick Jensen and John Zerzan among others. Issue two came in the following year in the wake of Sept. 11 and took a major step from the first issue in becoming something of its own rather than another mouthpiece of green anarchist rhetoric. The articles took a more in depth direction opening a more analytical and critical draw between anarchy and anthropology, attacks on Reason and the Progress/linear views of human history and Future that stand at the base of the ideology of civilization.

Vegan Anarchism

Veganarchism or vegan anarchism, is the political philosophy of veganism (more specifically animal liberation and earth liberation) and anarchism, creating a combined praxis that is designed to be a means for social revolution. This encompasses viewing the state as unnecessary and harmful to animals, both human and non-human, whilst practising a vegan lifestyle. It is either perceived as a combined theory, or that both philosophies are essentially the same. It is further described as an anti-speciesist perspective on green anarchism, or an anarchist perspective on animal liberation.

Veganarchists typically view oppressive dynamics within society to be interconnected, from statism, racism and sexism to human supremacy and redefine veganism as a radical philosophy that sees the state as harmful to animals. Those who believe in veganarchy can be either against reform for animals or for it, although do not limit goals to changes within the law.

Total Liberation

Total liberationism is a form of green anarchism that combines an opposition to all forms of human oppression with a commitment to animal and earth liberation. Whilst more conventional approaches to anarchist politics typically maintain a tacit assumption of anthropocentrism, proponents of total liberation espouse a holistic revolutionary strategy aimed at identifying the intersections between all forms of domination and social hierarchy, and building alliances between individual political movements in order to integrate them into a single movement aimed at abolishing a range of social structures such as the state, capitalism, patriarchy, racism, heterosexism, cissexism, disablism, ageism, speciesism, and ecological domination. As David Pellow summarises:

The concept of total liberation stems from a determination to understand and combat all forms of inequality and oppression. I propose that it comprises four pillars: (1) an ethic of justice and anti-oppression inclusive of humans, nonhuman animals, and ecosystems; (2) anarchism; (3) anti-capitalism; and (4) an embrace of direct action tactics.

Derrick Jensen

Derrick Jensen is an American author and environmental activist (and critic of mainstream environmentalism) living in Crescent City, California. Jensen's work is sometimes characterized as anarcho-primitivist, although he has categorically rejected that label, describing primitivist as a "racist way to describe indigenous peoples." He prefers to be called "indigenist" or an "ally to the indigenous," because "indigenous peoples have had the only sustainable human social organizations, and... we need to recognize that we [colonizers] are all living on stolen land."

A Language Older Than Words uses the lens of domestic violence to look at the larger violence of western culture. *The Culture of Make Believe* begins by exploring racism and misogyny and moves to examine how this culture's economic system leads inevitably to hatred and atrocity. *Strangely Like War* is about deforestation. *Walking on Water* is about education (It begins: "As is true for most people I know, I've always loved learning. As is also true for most people I know, I always hated school. Why is that?"). *Welcome to the Machine* is about surveillance, and more broadly about science and what he perceives to be a Western obsession with control. *Resistance Against Empire* consists of interviews with J. W. Smith (on poverty), Kevin Bales (on slavery), Anuradha Mittal (on hunger), Juliet Schor ('globalization' and environmental degradation), Ramsey Clark (on US 'defense'), Stephen Schwartz (editor of *The Nonproliferation Review*, on nukes), Alfred McCoy (politics and heroin), Christian Parenti (the US prison system), Katherine Albrecht (on RFID), and Robert McChesney (on (freedom of) the media) conducted between 1999 and 2004. *Endgame* is about what he describes as the inherent unsustainability of civilization. In this book he asks: "Do you believe that this culture will undergo a voluntary transformation to a sane and sustainable way of living?" Nearly everyone he talks to says no. His next question is: "How would this understanding – that this culture will not voluntarily stop destroying the natural world, eliminating indigenous cultures, exploiting the poor, and killing those who resist – shift our strategy and tactics? The answer? Nobody knows, because we never talk about it: we're too busy pretending the culture will undergo a magical transformation." *Endgame*, he says, is "about that shift in strategy, and in tactics." Jensen co-wrote the book *Deep Green Resistance: Strategy to Save the Planet* with Lierre Keith and Aric McBay.

CrimethInc.

CrimethInc. is a decentralized anarchist collective of autonomous cells. CrimethInc. emerged in the mid-1990s, initially as the hardcore zine *Inside Front*, and began operating as a collective in 1996. It has since published widely read articles and zines for the anarchist movement and distributed posters and books of its own publication. Individuals adopting the CrimethInc. *nom de guerre* have included convicted ELF arsonists, as well as hacktivists who successfully attacked the websites of DARE, Republican National Committee and sites related to U.S. President George W. Bush's 2004 re-election campaign. The creation of propaganda has been described as the collectives' core function. Among their best-known publications are the books *Days of War, Nights of Love, Expect Resistance, Evasion, Recipes for Disaster: An Anarchist Cookbook* and the pamphlet *Fighting For Our Lives* (of which, to date, they claim to have printed 600,000 copies), the hardcore punk/political zine *Inside Front*, and the music of hardcore punk bands. As well as the traditional anarchist opposition to the state and capitalism, agents have, at times, advocated a straight edge lifestyle, the total supersession of gender roles, violent insurrection against the state, and the refusal of work.

Direct Action

Some Green Anarchists engage in direct action. Organizing themselves through groups like

Earth First!, Root Force, or more drastically, the Earth Liberation Front (ELF), Earth Liberation Army (ELA) and Animal Liberation Front (ALF). They may take direct action against what they see as systems of oppression, such as the logging industry, the meat and dairy industries, animal testing laboratories, genetic engineering facilities and, more rarely, government institutions.

Such actions are usually, though not always, non-violent, with groups such as The Olga Cell attempting assassinations of nuclear scientists, and other related groups sending letterbombs to nano tech and nuclear tech-related targets. Though not necessarily Green anarchists, activists have used the names Animal Rights Militia, Justice Department and Revolutionary Cells among others, to claim responsibility for openly violent attacks.

Convictions

Rod Coronado is an eco-anarchist and is an unofficial spokesperson for the Animal Liberation Front and Earth Liberation Front. On February 28, 1992, Coronado carried out an arson attack on research facilities at Michigan State University (MSU), and released mink from a nearby research farm on campus, an action claimed by the ALF, and for which Coronado was subsequently convicted.

In 1997, the editors of *Green Anarchist* magazine and two British supporters of the Animal Liberation Front were tried in connection with conspiracy to incite violence, in what came to be known as the GANDALF trial.

Green anarchist Tre Arrow was sought by the FBI in connection with an ELF arson on April 15, 2001 at Ross Island Sand and Gravel in Portland, torching three trucks amounting of $200,000 in damage. Another arson occurred a month later at Ray Schoppert Logging Company in Estacada, Oregon, on June 1, 2001 against logging trucks and a front loader, resulting in $50,000 damage. Arrow was indicted by a federal grand jury in Oregon and charged with four felonies for this crime on October 18, 2002. On March 13, 2004, after fleeing to British Columbia, he was arrested in Victoria for stealing bolt cutters and was also charged with being in Canada illegally. He was then sentenced on August 12, 2008 to 78 months in federal prison for his part in the arson and conspiracy ELF attacks in 2001.

In January 2006, Eric McDavid, a green anarchist, was convicted of conspiring to use fire or explosives to damage corporate and government property. On March 8, he formally declared a hunger strike due to the jail refusing to provide him with vegan food. He has been given vegan food off and on since. In September 2007, he was convicted on all counts after the two activists he conspired with pleaded guilty testified against him. An FBI confidential source named “Anna” was revealed as a fourth participant, in what McDavid’s defense argued was entrapment. In May 2008, he was sentenced to nearly 20 years in prison.

On March 3, 2006, a federal jury in Trenton, New Jersey convicted six members of SHAC, including green-anarchist Joshua Harper, for “terrorism and Internet stalking”, according to the *New York Times*, finding them guilty of using their website to “incite attacks” on those who did business with Huntingdon Life Sciences HLS. In September 2006, the SHAC 7 received jail sentences of 3 to 6 years.

Other prisoners

- Marco Camenisch; Swiss green anarchist accused of arson against electricity pylon.

- Nicole Vosper; green anarchist who pleaded guilty to charges against HLS.
- Marius Mason, (born Marie Jeanette Mason) #04672-061, FMC Carswell, Federal Medical Center, P.O. Box 27137, Fort Worth, TX 76127, USA. Serving 21 years and 10 months for his involvement in an ELF arson against a University building carrying out Genetically Modified crop tests. Marius also pleaded guilty to conspiring to carry out ELF actions and admitted involvement in 12 other ELF actions. (vegan).

Green Conservatism

Green conservatism is a combination of conservatism with environmentalism. Environmental concern has been voiced by both conservative politicians and philosophers throughout the history of modern conservatism with Edmund Burke (the philosophical founder of modern conservatism) in his *Reflections on the Revolution in France* quoted as saying 'the earth, the kind and equal mother of all ought not to be monopolised to foster the pride and luxury of any men'.

Variants

Australia

Prime Minister Malcolm Fraser legislated significant environmental protection laws for the Great Barrier Reef while leader of the conservative Liberal Party as early as the 1970s.

Australian politics became more polemic around the year 2000. Once an uneasy alliance between the Australian Labor Party (the traditional opposition to the Liberal Party) and The Greens had formed, it caused many conservative voters to become suspicious of Green politics.

In 2004, the conservative Family First Party aired a television advertisement labelling The Australian Greens as the *Extreme Greens*. Since then, the continuously increasing anti-conservative dominance within The Greens has led to this label receiving broader use in the media.

Brazil

The Sustainability Network party was founded by Marina Silva, a prominent conservative environmentalist who was previously a member of various political parties including the Brazilian Green Party.

The National Ecologic Party also has ties to the Assemblies of God, the largest evangelical denomination present in Brazil.

Canada

In Canada, the term was popularized in 2006 by Preston Manning, former federal opposition leader and founder of the Reform Party of Canada. Specifically Manning has argued that Western Canadian Conservatism with its strong rural roots and populist rhetoric will eventually have to reconcile the need for economic growth with protection of the environment. He has specifically talked about using water pricing in the Athabasca Oil Sands to prevent a "tragedy of the commons" scenario.

Denmark

In Denmark, the Conservative People's Party advocates conservative policies, while being a part of the green bloc of Folketinget. They opposed the Danish 2016 Agriculture Reform due to enivornmental concerns.

Germany

In Germany, the Ecological Democratic Party was formed by more right-wing defectors from Die Grünen in 1982. It combined a focus on environmental protection with a promotion of the right to life (opposition to abortion, euthanasia and capital punishment); it differs from The Greens by being less supportive of immigration and restrictions on state powers in criminal justice issues, not focusing on gay and lesbian rights, and having a differing view on feminism.

While having never gained seats in federal or state legislatures in Germany, it made a name for itself by its involvement in the opposition to a Czech nuclear reactor in Temelin, across the border from Bavaria. It led an initiative for a popular referendum to abolish the Bavarian Senate (that state's upper house) which was successful. It is still active in the present day.

Japan

In Japan, the Environmental Green Political Assembly, or Midori no Kaigi, emerged from the conservative reformist Sakigake Party. It combined a conservative ideology with an ecologist platform, forcing out a number of non-ecologist members to join the Democratic Party's *Ryoun-kai* faction. It showed poor performance at the polls, and was dissolved in 2004, merged into the conservative Liberal Democratic Party.

Latvia

Following the 2014 Latvian parliamentary election, the Union of Greens and Farmers became the third largest party in the country with 21 seats.

Lithuania

In the 2016 Lithuanian parliamentary election, the Lithuanian Farmers and Greens Union became the largest party in a surprise victory with 51 seats.

Mexico

The Ecologist Green Party of Mexico, founded in 1993, won 47 seats in the 2015 Mexican legislative election.

United Kingdom

The Conservative Party under David Cameron promised a green agenda which included proposals designed to impose a tax on workplace car parking spaces, a halt to airport growth, a tax on petrol thirsty 4x4s and restrictions on car advertising. The measures were suggested by The Quality of Life Policy Group, which was set up by Cameron to help fight climate change.

Cameron spoke of embracing 'green' issues, and had made climate change a key component of his speeches. He called for an independent climate change commission to ensure that emissions reductions targets are met. However, Cameron's claim of leading the "greenest government ever" has been repudiated by Green Party MP Caroline Lucas, who argued that Cameron had "shown little interest in green policy and the sustainability agenda."

Zac Goldsmith, who was the Conservative candidate for London Mayor in 2016, describes himself as an environmentalist. He received the *Global Green Award for International Environmental Leadership* in 2004.

Bright Blue, a liberal conservative think tank launched the Green Conservatism project to create conservative support for policies that address climate change.

Green conservatism is also supported by some on the right of the Conservative Party, such as Richard Body, on the grounds that environmental conservation is consistent with the ideas of traditionalism and High Toryism.

United States

One of the first uses of the term *green conservatism* was by former United States Republican House Speaker Newt Gingrich, in a debate on environmental issues with John Kerry. Around this time, the green conservative movement was sometimes referred to as the *crunchy con* movement, a term popularized by *National Review* magazine and the writings of Rod Dreher.

In the United States, the Republican Party is generally considered as the conservative party. Green conservatism manifested itself as a movement in groups such as ConservAmerica, which seeks to strengthen the Republican Party's stance on environmental issues and support efforts to conserve natural resources and protect human and environmental health.

The Independent Greens of Virginia (or Indy Greens) call themselves "common sense conservatives". The party, over the last decade, has run many conservative greens for local, state, and federal office. In 2004, the party gave its ballot line to Constitution Party nominee Michael Peroutka for president, and in 2008, once again placed the Constitution Party nominee Chuck Baldwin on the ballot as its presidential candidate. The Indy Greens call for balancing the federal budget and paying off the federal debt.

The Republican Party had long supported the protection of the environment at the first half of 20th Century. Republican President Theodore Roosevelt was a prominent conservationist whose policies eventually led to the creation of the modern National Park Service. Republican President Richard Nixon was responsible for establishing the Environmental Protection Agency in 1970.

More recently, California Republican Governor Arnold Schwarzenegger, with the support of 16 other states, sued the Federal Government and the United States Environmental Protection Agency for the right to set vehicle emission standards higher than the Federal Standard, a right to which California is entitled under the Clean Air Act.

This association however has shifted as the Democratic Party came to also support environmentalism. For example, Democratic President Bill Clinton did not send the Kyoto Protocol to the U.S.

Senate for ratification, as he thought it unfair to the United States. President George W. Bush also publicly opposed ratification of the Kyoto Protocols on the grounds that they unfairly targeted Western industrialized nations such as the United States while favoring developing Global South polluters such as China and India.

In 2000, the Republican Party adopted as part of its platform support for the development of market-based solutions to environmental problems. According to the platform, "economic prosperity and environmental protection must advance together, environmental regulations should be based on science, the government's role should be to provide market-based incentives to develop the technologies to meet environmental standards, we should ensure that environmental policy meets the needs of localities, and environmental policy should focus on achieving results processes."

The Bush administration, along with several of the candidates that sought the Republican Presidential nomination in 2008, supported increased Federal investment into the development of clean alternative fuels, increased nuclear power, as well as fuels such as ethanol, as a way of helping the U.S. achieve energy independence, as opposed to supporting less use of carbon dioxide-producing methods of generating energy. John McCain, who ran unsuccessfully for President in 2008, supported the cap-and-trade policy, a policy that is quite popular among Democrats but much less so among other Republicans. Some Republicans support increased oil drilling in protected areas such as the Arctic National Wildlife Refuge, a position that has drawn sharp criticism from some activists. By the 2012 election the apparent support for action in the Republican Party had diminished, with eventual nominee Mitt Romney stating his opposition to cap-and-trade and voicing scepticism of links between global warming and human activity.

Green Liberalism

Green liberalism or liberal environmentalism is liberalism that includes green politics into its ideology. Green liberals are usually liberal on social issues and "green" on economic issues.

Philosophy

Green liberalism values the Earth very highly, and this philosophy highly values the planet being passed down to the next generation unharmed. Green liberalism accepts that the natural world is a system in a state of flux, and does not seek to conserve the natural world as it is. However, it does seek to minimize the damage done by the human species on the natural world, and to aid the regeneration of damaged areas.

In economic issues, green liberals take a position somewhere between classical liberalism (on the center/center-right) and social liberalism (on the center/center-left): they may favor slightly less government involvement than social liberals, but far more than classical liberals. Some within the circle of green liberals practice free-market environmentalism and thus, sharing similarity with rightist classical liberalism or libertarianism. This is one of a few reasons why a blue-green alliance is possible in politics.

The historian Conrad Russell, a British Liberal Democrat member of the House of Lords, dedicated a chapter of his book *The Intelligent Person's Guide to Liberalism* to the subject of green liberalism. The term "green liberalism" was coined, however, by political philosopher Marcel Wissenburg in – among others – his 1998 book *Green Liberalism: The free and the green society*.

The Liberal Party of Canada under Stéphane Dion placed the environment at the front of its political agenda, proposing an ecotax and tax shift it called the Green Shift. Similarly, the Liberal Democrats (UK) have drawn on the same concept to propose a "Green Tax Switch".

Green Libertarianism

Green libertarianism (also known as eco-libertarianism) is a hybrid political philosophy that has developed in the United States. Based upon a mixture of political third party values, such as the environmental and economic platform from the Green Party of the United States and the civil liberties platform of the U.S. Libertarian Party, the green libertarian philosophy attempts to consolidate progressive or agrarian values with libertarianism. While green libertarians have tended to associate with the U.S. Green Party, the movement has grown to encompass economic liberals who advocate free markets and commonly identify with contemporary American libertarianism.

History

Left-libertarian political philosophy, like that of the greens, is historically rooted in the individualist and social schools of anarchism. Anarcho-communist theorist Peter Kropotkin, a Russian prince and leading opponent of *laissez-faire* Social Darwinism, developed a theory of how "mutual aid" is the real basis for social organization in his *Mutual Aid: A Factor in Evolution*. Murray Bookchin and the Institute for Social Ecology sought to further elaborate these ideas. Bookchin was one of the main influences behind the formation of the German Green Party, the first green party to win seats in state and national parliaments.

Some more moderate, green libertarians are both egalitarian and democratic. New England Transcendentalism (especially Henry David Thoreau and Bronson Alcott) and German Romanticism, the Pre-Raphaelites, and other "back to nature" movements combined with anti-war, anti-industrialism, civil liberties, and decentralization movements are all part of this tradition. The modern Green Party of the United States seeks to apply these ideas to a more pragmatic system of democratic governance, as opposed to contemporary right or left anarchism.

"Natural Capitalism"

Sustainability advocates Paul Hawken, Amory Lovins and Hunter Lovins posited in their book *Natural Capitalism: Creating the Next Industrial Revolution*, published in 2000, that elements of libertarianism and green politics could be coalesced to produce economic as well as environmental benefits. The 2006 book, *Green to Gold*, written by environmental scholars Daniel C. Esty and Andrew S. Winston provided ideas on how companies can apply the principles of green libertarianism.

The work of Austrian school economist Friedrich Hayek is especially important to understanding the organic view of society and how most human institutions, including law and the economy, are "the result of human action but not of human design." In his last major work, *Law, Legislation and Liberty*, Hayek differentiates between endogenous orders, or self-organizing systems, and exogenous orders imposed from without.

Hayek argues that free and sustainable societies and economies which support them should follow general guidelines, such as "7th Generation Principle" rather than specific economic laws and regulations.

Balance of Ecology and Economics

A fundamental concern among green libertarians is the health of global ecology and carrying capacity in view of climate adaptation. The green libertarian philosophy recognizes that ecology and economics are inseparable. It seeks a system of effective environmental law that is compatible with civil liberties and market economy.

Green libertarians believe there should be a clear distinction between science and political ideology. For example, a green libertarian might be concerned by the phrases such as "wealth redistribution" and "reducing poverty" in the Stern Review and in some IPCC documents and statements. Among green libertarians, the preservation of civil and economic individual freedom may take precedence over long-term climate concerns because, ultimately, humans are part of nature. They believe that natural ecologies, like the free markets, are dynamic and self-adjusting systems.

Limited Government

As part of the libertarian tradition, green libertarians maintain that the government itself is responsible for most environmental degradation, either directly, or by encouraging and protecting politically powerful corporations and other organized interests which degrade, pollute and deplete the natural environment. Therefore, the government should be held accountable to all the same environmental regulations they place on businesses. One problem is that while private corporations or individuals can be sued under the Common Law for damaging the environment, the government protects itself from the same suits. Therefore, green libertarians call for the abolition of sovereign immunity. Increasingly, federal and state law is being amended by lobbyists for those who pollute or extract resources from public lands or waterways so that such actions can no longer be challenged in the courts.

The green libertarian philosophy supports constitutionally limited government, grassroots democracy, and decentralized minarchism. Although many in the movement oppose government regulation of business, believing it to be generally counterproductive, they contend that different legal and economic principles such as full-cost accounting or "internalizing externalities" – rather than government regulations – would be more effective at remedying problems such as pollution. A central tenet of a libertarian environmentalist stance is that corporate externalities are not priced into the market correctly, creating market distortions in the valuation and price of goods, healthy living and the value of the environment. Greenhouse gases should be taxed directly, according to a formula which calculates the negative costs to the global environment of burning more non-renewable fossil fuels. This approach, it is argued, also has the advantage of providing the correct price signals to utilities and other energy consumers so that they can rapidly convert to more environmentally friendly technologies. The Rocky Mountain Institute advocates this kind of market-based environmental protection strategy.

References

- Ruether, Rosemary Radford (2003). Heather Eaton & Lois Ann Lorentzen, ed. Ecofeminism and Globalization. Lanham, Boulder, New York, Toronto, Oxford: Rowman and Littlefield. pp. vii – xi. ISBN 0-7425-2697-6
- Bernstein, Emily (1993). "Neighborhood Report: Harlem; Sowing a Future With Green in It". The New York Times. Retrieved October 9, 2016
- Paul Aron; Luke Miller (2007). "The Third Team: A brief history of the Australian Democrats after 30 years" (PDF). Australian Democrats. Retrieved 2009-04-05
- Gaard, Greta (2011). "Ecofeminism Revisited: Rejecting Essentialism and Re-Placing Species in a Material Feminist Environmentalism". Feminist Formations. 23 (2): 26–53. doi:10.1353/ff.2011.0017
- MacGregor, Sherilyn (2006). Beyond mothering earth: ecological citizenship and the politics of care. Vancouver: UBC Press. p. 286. ISBN 0-7748-1201-X
- Wenzer, Kenneth C. (1997). "Tolstoy's Georgist Spiritual Political Economy (1897–1910): Anarchism and Land Reform". The American Journal of Economics and Sociology. 56 (4, Oct). JSTOR 3487337
- "History of Vegetarianism – Élisée Reclus (1830 – 1905)". ivu.org. International Vegetarian Union. Retrieved January 23, 2010
- BERG, JOHN C. (1 March 2003). "Reviews Editor's Introduction". New Political Science. 25 (1): 129–143. doi:10.1080/0739314032000071262 – via IngentaConnect
- Warren, Karen J. (2000). Ecofeminist Philosophy: A Western Perspective on What It Is and Why It Matters. Lanham, Maryland: Roman & Littlefield Publishers, Inc. ISBN 9780847692996
- Marks, Gary & Wilson, Carole (July 2000). "The Past in the Present: A Cleavage Theory of Party Response to European Integration" (PDF). British Journal of Political Science. 30: 433–459. doi:10.1017/S0007123400000181
- Power, Matthew (March 2008). "Mississippi Drift" (PDF). Harper's Magazine: 54–63. Archived from the original (PDF) on 2008-05-11. Retrieved 2008-03-10
- Mann, Susan A (2011). "Pioneers of U.S. Ecofeminism and Environmental Justice". Feminist Formations. 23 (2): 1–25. doi:10.1353/ff.2011.0028
- Vakoch, Douglas A (2011-01-01). Ecofeminism and rhetoric: critical perspectives on sex, technology, and discourse. New York: Berghahn Books. ISBN 9780857451873
- Michael J. Prince, "Canadian disability activism and political ideas: In and between neo-liberalism and social liberalism." Canadian Journal of Disability Studies 1.1 (2012): 1-34
- Bush, George W. (2001-03-13). "Text of a Letter from the President". Archived from the original on July 22, 2009. Retrieved 2007-11-09
- Vakoch, Douglas A (2012-01-01). Feminist ecocriticism: environment, women, and literature. Lanham, Md.: Lexington Books. ISBN 9780739176825
- Fotopoulos, Takis (October 2004). "Why an Inclusive Democracy? The multidimensional crisis, globalisation and inclusive democracy". The International Journal of Inclusive Democracy. 1 (1). Retrieved 2008-05-21
- Bookchin, Murray (11 July 1992). "Deep Ecology, Anarcho-Syndicalism and the Future of Anarchist Thought". Institute for Social Ecology. Retrieved 5 November 2014

3

Philosophies in Green Politics

Agrarianism is a philosophy which considers rural society to be better than urban society. It believes in promoting the simple rural life as opposed to the life people live in big cities. This chapter has been carefully written to provide an easy understanding of the varied facets of green politics.

Agrarianism

Agrarianism is a social philosophy or political philosophy which values rural society as superior to urban society, the independent farmer as superior to the paid worker, and sees farming as a way of life that can shape the ideal social values. It stresses the superiority of a simpler rural life as opposed to the complexity of city life.

Philosophy

M. Thomas Inge defines agrarianism by the following basic tenets:

- Farming is the sole occupation that offers total independence and self-sufficiency.
- Urban life, capitalism, and technology destroy independence and dignity and foster vice and weakness.
- The agricultural community, with its fellowship of labor and co-operation, is the model society.
- The farmer has a solid, stable position in the world order. He "has a sense of identity, a sense of historical and religious tradition, a feeling of belonging to a concrete family, place, and region, which are psychologically and culturally beneficial." The harmony of his life checks the encroachments of a fragmented, alienated modern society.
- Cultivation of the soil "has within it a positive spiritual good" and from it the cultivator acquires the virtues of "honor, manliness, self-reliance, courage, moral integrity, and hospitality." They result from a direct contact with nature and, through nature, a closer relationship to God. The agrarian is blessed in that he follows the example of God in creating order out of chaos.

History

The philosophical roots of agrarianism include European and Chinese philosophers. The Chinese school of Agriculturalism was a philosophy that advocated peasant utopian communalism and egalitarianism. In societies influenced by Confucianism, the farmer was considered an esteemed productive member of society, but merchants who made money were looked down upon. That influenced European intellectuals like François Quesnay, an avid Confucianist and advocate of China's agrarian policies, in forming the French agrarian philosophy of physiocracy. The physiocrats, along with the ideas of John Locke and the Romantic Era, formed the basis of modern European and American agrarianism.

United States president (1801–1809) Thomas Jefferson was a representative agrarian who built Jeffersonian democracy around the notion that farmers are "the most valuable citizens" and the truest republicans.

Agrarian Parties

Peasant parties first appeared across Eastern Europe between 1860 and 1910, when commercialized agriculture and world market forces disrupted traditional rural society, and the railway and growing literacy facilitated the work of roving organizers. Agrarian parties advocated land reforms to redistribute land on large estates among those who work it. They also wanted village cooperatives to keep the profit from crop sales in local hands and credit institutions to underwrite needed improvements. Many peasant parties were also nationalist parties because peasants often worked their land for the benefit of landlords of different ethnicity.

Peasant parties rarely had any power before World War I but some became influential in the inter-war era, especially in Bulgaria and Czechoslovakia. For a while, in the 1920s and the 1930s, there was a Green International (International Agrarian Bureau) based on the peasant parties in Bulgaria, Czechoslovakia, Poland, and Serbia. It functioned primarily as an information center that spread the ideas of agrarianism and combating socialism on the left and landlords on the right and never launched any significant activities.

Africa

Tunisia

The Farmers' Voice Party won a seat in the district of Jendouba after the parliamentary election of 2014.

Europe

Bulgaria

In Bulgaria, the Bulgarian Agrarian National Union (BZNS) was organized in 1899 to resist taxes and build cooperatives. BZNS came to power in 1919 and introduced many economic, social, and legal reforms. However, conservative forces crushed BZNS in a 1923 coup and assassinated its leader, Aleksandar Stamboliyski (1879–1923). BZNS was made into a communist puppet group until 1989, when it reorganized as a genuine party.

Czechoslovakia

In Czechoslovakia, the Republican Party of Agricultural and Smallholder People often shared power in parliament as a partner in the five-party pětka coalition. The party's leader, Antonin Svehla

(1873–1933), was prime minister several times. It was consistently the strongest party, forming and dominating coalitions. It moved beyond its original agrarian base to reach middle-class voters. The party was banned by the National Front after the Second World War.

France

In France, the Hunting, Fishing, Nature, Tradition party is a moderate conservative, agrarianist party, reaching a peak of 4,23% in the French presidential election, 2002. It would later on become affiliated to France's main conservative party, Union for a Popular Movement.

Latvia

In Latvia, the Union of Greens and Farmers is supportive of traditional small farms and perceives them as more environmentally friendly than large-scale farming: Nature is threatened by development, while small farms are threatened by large industrial-scale farms.

Lithuania

In Lithuania, as of 2017, the government is led by the Lithuanian Farmers and Greens Union, under the leadership of industrial farmer Ramūnas Karbauskis.

Poland

In Poland, the Polish People's Party traces its tradition to an agrarian party in Austro-Hungarian-controlled Galician Poland. After the fall of a communist regime, PPP biggest success came in 1993 elections, where it won 132 out of 460 parliamentary seats. Since then, PPP's support steadly declines.

Romania

In Romania, older parties from Transylvania, Moldavia, and Wallachia merged to become the National Peasants' Party in 1926. Iuliu Maniu (1873–1953) was a prime minister with an agrarian cabinet from 1928–1930 and briefly in 1932–1933, but the Great Depression made proposed reforms impossible. The communist regime dissolved the party in 1947, but it reformed in 1989 after they fell from power.

The reformed party, which also incorporated elements of Christian democracy in its ideology, governed Romania as part of the Romanian Democratic Convention between 1996–2000.

Serbia

In Serbia, Nikola Pašić (1845–1926) and his People's Radical Party dominated Serbian politics after 1903. The party also monopolized power in Yugoslavia from 1918 to 1929. During the dictatorship of the 1930s, the prime minister was from that party.

Ukraine

In Ukraine, the Radical Party of Oleh Lyashko has promised to purify the country of oligarchs "with a pitchfork". The party advocates a number of traditional left-wing positions (lower salary taxes, a ban on agricultural land sale and eliminating the illegal land market, a tenfold increase in budget spending on health, setting up primary health centres in every village), and mixes them with strong nationalist sentiments.

Oceania

Australia

Historian F.K. Crowley finds that:

> Australian farmers and their spokesman have always considered that life on the land is inherently more virtuous, as well as more healthy, more important and more productive, than life in the towns and cities. The farmers complained that something was wrong with an electoral system which produced parliamentarians who spent money beautifying vampire-cities instead of developing the interior.

The National Party of Australia (formerly called the Country Party), from the 1920s to the 1970s, promulgated its version of agrarianism, which it called "countrymindedness". The goal was to enhance the status of the graziers (operators of big sheep ranches) and small farmers and justified subsidies for them.

New Zealand

The New Zealand Liberal Party aggressively promoted agrarianism in its heyday (1891–1912). The landed gentry and aristocracy ruled Britain at this time. New Zealand never had an aristocracy but its wealthy landowners largely controlled politics before 1891. The Liberal Party set out to change that by a policy it called "populism." Richard Seddon had proclaimed the goal as early as 1884: "It is the rich and the poor; it is the wealthy and the landowners against the middle and labouring classes. That, Sir, shows the real political position of New Zealand." The Liberal strategy was to create a large class of small landowning farmers who supported Liberal ideals. The Liberal government also established the basis of the later welfare state such as old age pensions and developed a system for settling industrial disputes, which was accepted by both employers and trade unions. In 1893, it extended voting rights to women, making New Zealand the first country in the world to do so.

To obtain land for farmers, the Liberal government from 1891 to 1911 purchased 3,100,000 acres (1,300,000 ha) of Maori land. The government also purchased 1,300,000 acres (530,000 ha) from large estate holders for subdivision and closer settlement by small farmers. The Advances to Settlers Act (1894) provided low-interest mortgages, and the agriculture department disseminated information on the best farming methods. The Liberals proclaimed success in forging an egalitarian, anti-monopoly land policy. The policy built up support for the Liberal Party in rural North Island electorates. By 1903, the Liberals were so dominant that there was no longer an organized opposition in Parliament.

Back-to-the-land Movement

Agrarianism is similar to but not identical with the back-to-the-land movement. Agrarianism concentrates on the fundamental goods of the earth, on communities of more limited economic and political scale than in modern society, and on simple living, even when the shift involves questioning the "progressive" character of some recent social and economic developments. Thus, agrarianism is not industrial farming, with its specialization on products and industrial scale.

Environmentalism

Environmentalism or environmental rights is a broad philosophy, ideology, and social movement regarding concerns for environmental protection and improvement of the health of the environment, particularly as the measure for this health seeks to incorporate the concerns of non-human elements. While environmentalism focuses more on the environmental and nature-related aspects of green ideology and politics, ecologism combines the ideology of social ecology and environmentalism. 'Ecologism' is more commonly used in continental European languages while 'environmentalism' is more commonly used in English but the words have slightly different connotations.

Environmentalism advocates the lawful preservation, restoration and/or improvement of the natural environment, and may be referred to as a movement to control pollution or protect plant and animal diversity. For this reason, concepts such as a land ethic, environmental ethics, biodiversity, ecology, and the biophilia hypothesis figure predominantly.

At its crux, environmentalism is an attempt to balance relations between humans and the various natural systems on which they depend in such a way that all the components are accorded a proper degree of sustainability. The exact measures and outcomes of this balance is controversial and there are many different ways for environmental concerns to be expressed in practice. Environmentalism and environmental concerns are often represented by the color green, but this association has been appropriated by the marketing industries for the tactic known as greenwashing.

Environmentalism is opposed by anti-environmentalism, which says that the Earth is less fragile than some environmentalists maintain, and portrays environmentalism as overreacting to the human contribution to climate change or opposing human advancement.

Definitions

Environmentalism denotes a social movement that seeks to influence the political process by lobbying, activism, and education in order to protect natural resources and ecosystems. The word was first coined in 1922.

An *environmentalist* is a person who may speak out about our natural environment and the sustainable management of its resources through changes in public policy or individual behavior. This may include supporting practices such as informed consumption, conservation initiatives, investment in renewable resources, improved efficiencies in the materials economy, transitioning to new accounting paradigms such as Ecological economics and renewing and revitalizing our connections with non-human life.

In various ways (for example, grassroots activism and protests), environmentalists and environmental organizations seek to give the natural world a stronger voice in human affairs.

In general terms, environmentalists advocate the sustainable management of resources, and the protection (and restoration, when necessary) of the natural environment through changes in public policy and individual behavior. In its recognition of humanity as a participant in ecosystems, the movement is centered around ecology, health, and human rights.

History

A concern for environmental protection has recurred in diverse forms, in different parts of the world, throughout history. For example, in Europe, King Edward I of England banned the burning of sea-coal by proclamation in London in 1272, after its smoke had become a problem. The fuel was so common in England that this earliest of names for it was acquired because it could be carted away from some shores by the wheelbarrow.

Earlier in the Middle East, the Caliph Abu Bakr in the 630s commanded his army to "Bring no harm to the trees, nor burn them with fire," and "Slay not any of the enemy's flock, save for your food." Arabic medical treatises during the 9th to 13th centuries dealing with environmentalism and environmental science, including pollution, were written by Al-Kindi, Qusta ibn Luqa, Al-Razi, Ibn Al-Jazzar, al-Tamimi, al-Masihi, Avicenna, Ali ibn Ridwan, Ibn Jumay, Isaac Israeli ben Solomon, Abd-el-latif, Ibn al-Quff, and Ibn al-Nafis. Their works covered a number of subjects related to pollution, such as air pollution, water pollution, soil contamination, municipal solid waste mishandling, and environmental impact assessments of certain localities.

Early Environmental Legislation

Levels of air pollution rose during the Industrial Revolution, sparking the first modern environmental laws to be passed in the mid-19th century.

At the advent of steam and electricity the muse of history holds her nose and shuts her eyes (H. G. Wells 1918).

The origins of the environmental movement lay in the response to increasing levels of smoke pollution in the atmosphere during the Industrial Revolution. The emergence of great factories and the concomitant immense growth in coal consumption gave rise to an unprecedented level of air pollution in industrial centers; after 1900 the large volume of industrial chemical discharges added to the growing load of untreated human waste. The first large-scale, modern environmental laws came in the form of Britain's Alkali Acts, passed in 1863, to regulate the deleterious air pollution (gaseous hydrochloric acid) given off by the Leblanc process, used to produce soda ash. An Alkali inspector and four sub-inspectors were appointed to curb this pollution. The responsibilities of the inspectorate were gradually expanded, culminating in the Alkali Order 1958 which placed all major heavy industries that emitted smoke, grit, dust and fumes under supervision.

In industrial cities local experts and reformers, especially after 1890, took the lead in identifying environmental degradation and pollution, and initiating grass-roots movements to demand and achieve reforms. Typically the highest priority went to water and air pollution. The Coal Smoke Abatement Society was formed in 1898 making it one of the oldest environmental NGOs. It was founded by artist Sir William Blake Richmond, frustrated with the pall cast by coal smoke. Although there were earlier pieces of legislation, the Public Health Act 1875 required all furnaces and fireplaces to consume their own smoke. It also provided for sanctions against factories that emitted large amounts of black smoke. The provisions of this law were extended in 1926 with the Smoke Abatement Act to include other emissions, such as soot, ash and gritty particles and to empower local authorities to impose their own regulations.

It was, however, only under the impetus of the Great Smog of 1952 in London, which almost brought the city to a standstill and may have caused upward of 6,000 deaths that the Clean Air Act 1956 was passed and pollution in the city was finally brought to an end. Financial incentives were offered to householders to replace open coal fires with alternatives (such as installing gas fires), or for those who preferred, to burn coke instead (a byproduct of town gas production) which produces minimal smoke. 'Smoke control areas' were introduced in some towns and cities in which only smokeless fuels could be burnt and power stations were relocated away from cities. The act formed an important impetus to modern environmentalism, and caused a rethinking of the dangers of environmental degradation to people's quality of life.

The late 19th century also saw the passage of the first wildlife conservation laws. The zoologist Alfred Newton published a series of investigations into the *Desirability of establishing a 'Close-time' for the preservation of indigenous animals* between 1872 and 1903. His advocacy for legislation to protect animals from hunting during the mating season led to the formation of the Royal Society for the Protection of Birds and influenced the passage of the Sea Birds Preservation Act in 1869 as the first nature protection law in the world.

First Environmental Movements

Early interest in the environment was a feature of the Romantic movement in the early 19th century. The poet William Wordsworth travelled extensively in the Lake District and wrote that it is a "sort of national property in which every man has a right and interest who has an eye to perceive and a heart to enjoy".

Systematic efforts on behalf of the environment only began in the late 19th century; it grew out of the amenity movement in Britain in the 1870s, which was a reaction to industrialization, the growth of cities, and worsening air and water pollution. Starting with the formation of the Commons Preservation Society in 1865, the movement championed rural preservation against the encroachments of industrialisation. Robert Hunter, solicitor for the society, worked with Hardwicke

Rawnsley, Octavia Hill, and John Ruskin to lead a successful campaign to prevent the construction of railways to carry slate from the quarries, which would have ruined the unspoilt valleys of Newlands and Ennerdale. This success led to the formation of the Lake District Defence Society (later to become The Friends of the Lake District).

John Ruskin an influential thinker who articulated the Romantic ideal of environmental protection and conservation.

In 1893 Hill, Hunter and Rawnsley agreed to set up a national body to coordinate environmental conservation efforts across the country; the "National Trust for Places of Historic Interest or Natural Beauty" was formally inaugurated in 1894. The organisation obtained secure footing through the 1907 National Trust Bill, which gave the trust the status of a statutory corporation. and the bill was passed in August 1907.

An early "Back-to-Nature" movement, which anticipated the romantic ideal of modern environmentalism, was advocated by intellectuals such as John Ruskin, William Morris, George Bernard Shaw and Edward Carpenter, who were all against consumerism, pollution and other activities that were harmful to the natural world. The movement was a reaction to the urban conditions of the industrial towns, where sanitation was awful, pollution levels intolerable and housing terribly cramped. Idealists championed the rural life as a mythical Utopia and advocated a return to it. John Ruskin argued that people should return to a *small piece of English ground, beautiful, peaceful, and fruitful. We will have no steam engines upon it we will have plenty of flowers and vegetables we will have some music and poetry; the children will learn to dance to it and sing it.*

Practical ventures in the establishment of small cooperative farms were even attempted and old rural traditions, without the "taint of manufacture or the canker of artificiality", were enthusiastically revived, including the Morris dance and the maypole.

These ideas also inspired various environmental groups in the UK, such as the Royal Society for the Protection of Birds, established in 1889 by Emily Williamson as a protest group to campaign for greater protection for the indigenous birds of the island. The Society attracted growing support from the suburban middle-classes as well as support from many other influential figures, such as the ornithologist Professor Alfred Newton. By 1900, public support for the organisation had grown, and it had over 25,000 members. The Garden city movement incorporated many environmental concerns into its urban planning manifesto; the Socialist League and The Clarion movement also began to advocate measures of nature conservation.

Original title page of *Walden* by Henry David Thoreau.

The movement in the United States began in the late 19th century, out of concerns for protecting the natural resources of the West, with individuals such as John Muir and Henry David Thoreau making key philosophical contributions. Thoreau was interested in peoples' relationship with nature and studied this by living close to nature in a simple life. He published his experiences in the book *Walden*, which argues that people should become intimately close with nature. Muir came to believe in nature's inherent right, especially after spending time hiking in Yosemite Valley and studying both the ecology and geology. He successfully lobbied congress to form Yosemite National Park and went on to set up the Sierra Club in 1892. The conservationist principles as well as the belief in an inherent right of nature were to become the bedrock of modern environmentalism.

In the 20th century, environmental ideas continued to grow in popularity and recognition. Efforts were starting to be made to save some wildlife, particularly the American bison. The death of the last passenger pigeon as well as the endangerment of the American bison helped to focus the minds of conservationists and popularize their concerns. In 1916 the National Park Service was founded by US President Woodrow Wilson.

The Forestry Commission was set up in 1919 in Britain to increase the amount of woodland in Britain by buying land for afforestation and reforestation. The commission was also tasked with promoting forestry and the production of timber for trade. During the 1920s the Commission focused on acquiring land to begin planting out new forests; much of the land was previously used for agricultural purposes. By 1939 the Forestry Commission was the largest landowner in Britain.

During the 1930s the Nazis had elements that were supportive of animal rights, zoos and wildlife, and took several measures to ensure their protection. In 1933 the government created a stringent animal-protection law and in 1934, *Das Reichsjagdgesetz* (The Reich Hunting Law) was enacted which limited hunting. Several Nazis were environmentalists (notably Rudolf Hess), and species protection and animal welfare were significant issues in the regime. In 1935, the regime enacted the "Reich Nature Protection Act" (*Reichsnaturschutzgesetz*). The concept of the *Dauerwald* (best translated as the "perpetual forest") which included concepts such as forest management and protection was promoted and efforts were also made to curb air pollution.

In 1949, *A Sand County Almanac* by Aldo Leopold was published. It explained Leopold's belief that humankind should have moral respect for the environment and that it is unethical to harm it. The book is sometimes called the most influential book on conservation.

Throughout the 1950s, 1960s, 1970s and beyond, photography was used to enhance public awareness of the need for protecting land and recruiting members to environmental organizations. David Brower, Ansel Adams and Nancy Newhall created the Sierra Club Exhibit Format Series, which helped raise public environmental awareness and brought a rapidly increasing flood of new members to the Sierra Club and to the environmental movement in general. "This Is Dinosaur" edited by Wallace Stegner with photographs by Martin Litton and Philip Hyde prevented the building of dams within Dinosaur National Monument by becoming part of a new kind of activism called environmentalism that combined the conservationist ideals of Thoreau, Leopold and Muir with hard-hitting advertising, lobbying, book distribution, letter writing campaigns, and more. The powerful use of photography in addition to the written word for conservation dated back to the creation of Yosemite National Park, when photographs persuaded Abraham Lincoln to preserve the beautiful glacier carved landscape for all time. The Sierra Club Exhibit Format Series galvanized public opposition to building dams in the Grand Canyon and protected many other national treasures. The Sierra Club often led a coalition of many environmental groups including the Wilderness Society and many others. After a focus on preserving wilderness in the 1950s and 1960s, the Sierra Club and other groups broadened their focus to include such issues as air and water pollution, population concern, and curbing the exploitation of natural resources.

Post-war Expansion

In 1962, *Silent Spring* by American biologist Rachel Carson was published. The book cataloged the environmental impacts of the indiscriminate spraying of DDT in the US and questioned the logic of releasing large amounts of chemicals into the environment without fully understanding their effects on ecology or human health. The book suggested that DDT and other pesticides may cause cancer and that their agricultural use was a threat to wildlife, particularly birds. The resulting public concern led to the creation of the United States Environmental Protection Agency in 1970 which subsequently banned the agricultural use of DDT in the US in 1972. The limited use of DDT in disease vector control continues to this day in certain parts of the world and remains controversial. The book's legacy was to produce a far greater awareness of environmental issues and interest into how people affect the environment. With this new interest in environment came interest in problems such as air pollution and petroleum spills, and environmental interest grew. New pressure groups formed, notably Greenpeace and Friends of the Earth (US), as well as notable local organizations such as the Wyoming Outdoor Council, which was founded in 1967.

In the 1970s, the environmental movement gained rapid speed around the world as a productive outgrowth of the counterculture movement.

The world's first political parties to campaign on a predominantly environmental platform were the United Tasmania Group Tasmania, Australia and the Values Party of New Zealand. The first green party in Europe was the Popular Movement for the Environment, founded in 1972 in the Swiss canton of Neuchâtel. The first national green party in Europe was PEOPLE, founded in Britain in February 1973, which eventually turned into the Ecology Party, and then the Green Party.

Protection of the environment also became important in the developing world; the Chipko movement was formed in India under the influence of Mohandas Gandhi and they set up peaceful resistance to deforestation by literally hugging trees (leading to the term "tree huggers"). Their peaceful methods of protest and slogan "ecology is permanent economy" were very influential.

Another milestone in the movement was the creation of an Earth Day. Earth Day was first observed in San Francisco and other cities on March 21, 1970, the first day of spring. It was created to give awareness to environmental issues. On March 21, 1971, United Nations Secretary-General U Thant spoke of a spaceship Earth on Earth Day, hereby referring to the ecosystem services the earth supplies to us, and hence our obligation to protect it (and with it, ourselves). Earth Day is now coordinated globally by the Earth Day Network, and is celebrated in more than 175 countries every year.

The UN's first major conference on international environmental issues, the United Nations Conference on the Human Environment (also known as the Stockholm Conference), was held on June 5–16, 1972. It marked a turning point in the development of international environmental politics.

By the mid-1970s, many felt that people were on the edge of environmental catastrophe. The Back-to-the-land movement started to form and ideas of environmental ethics joined with anti-Vietnam War sentiments and other political issues. These individuals lived outside normal society and started to take on some of the more radical environmental theories such as deep ecology. Around this time more mainstream environmentalism was starting to show force with the signing of the Endangered Species Act in 1973 and the formation of CITES in 1975. Significant amendments were also enacted to the United States Clean Air Act and Clean Water Act.

In 1979, James Lovelock, a British scientist, published *Gaia: A new look at life on Earth*, which put forth the Gaia hypothesis; it proposes that life on earth can be understood as a single organism. This became an important part of the Deep Green ideology. Throughout the rest of the history of environmentalism there has been debate and argument between more radical followers of this Deep Green ideology and more mainstream environmentalists.

21st Century

Environmentalism continues to evolve to face up to new issues such as global warming, overpopulation and genetic engineering.

Research demonstrates a precipitous decline in the US public's interest in 19 different areas of environmental concern. Americans are less likely be actively participating in an environmental movement or organization and more likely to identify as "unsympathetic" to an environmental movement then in 2000. This is likely a lingering factor of the Great Recession in 2008. Since 2005, the percentage of Americans agreeing that the environment should be given priority over economic growth has dropped 10 points, in contrast, those feeling that growth should be given priority "even if the environment suffers to some extent" has risen 12 percent. These numbers point to the growing complexity of environmentalism and its relationship to economics.

New Forms of Ecoactivism

Tree sitting is a form of activism in which the protester sits in a tree in an attempt to stop the removal of a tree or to impede the demolition of an area with the longest and most famous tree-sitter being Julia Butterfly Hill, who spent 738 days in a California Redwood, saving a three-acre tract of forest.

Sit-in is a form of activism where one or any number of people occupy a place as a form of protest. The tactic can be used to encourage social change, such as the Greensboro sit-ins, a series of protests in 1960 to stop racial segregation, but can also be used in ecoactivism, as in the Dakota Access Pipeline Protest.

Environmental Movement

Before flue-gas desulfurization was installed, the air-polluting emissions from this power plant in New Mexico contained excessive amounts of sulfur dioxide.

The *environmental movement* (a term that sometimes includes the conservation and green movements) is a diverse scientific, social, and political movement. Though the movement is represented by a range of organizations, because of the inclusion of environmentalism in the classroom curriculum, the environmental movement has a younger demographic than is common in other social movements.

Environmentalism as a movement covers broad areas of institutional oppression, including for example: consumption of ecosystems and natural resources into waste, dumping waste into disadvantaged communities, air pollution, water pollution, weak infrastructure, exposure of organic life to toxins, mono-culture, anti-polythene drive (jhola movement) and various other focuses. Because of these divisions, the environmental movement can be categorized into these primary focuses: environmental science, environmental activism, environmental advocacy, and environmental justice.

Free Market Environmentalism

Free market environmentalism is a theory that argues that the free market, property rights, and tort law provide the best tools to preserve the health and sustainability of the environment. It considers environmental stewardship to be natural, as well as the expulsion of polluters and other aggressors through individual and class action.

Evangelical Environmentalism

Evangelical environmentalism is an environmental movement in the United States in which some Evangelicals have emphasized biblical mandates concerning humanity's role as steward and subsequent responsibility for the caretaking of Creation. While the movement has focused on different environmental issues, it is best known for its focus of addressing climate action from a biblically grounded theological perspective.

Preservation and Conservation

Environmental preservation in the United States and other parts of the world, including Australia, is viewed as the setting aside of natural resources to prevent damage caused by contact with humans or by certain human activities, such as logging, mining, hunting, and fishing, often to replace them with new human activities such as tourism and recreation. Regulations and laws may be enacted for the preservation of natural resources.

Organizations and Conferences

Reef doctor work station in Ifaty, Madagascar

Environmental organizations can be global, regional, national or local; they can be government-run or private (NGO). Environmentalist activity exists in almost every country. Moreover, groups dedicated to community development and social justice also focus on environmental concerns.

Some US environmental organizations, among them the Natural Resources Defense Council and the Environmental Defense Fund, specialize in bringing lawsuits (a tactic seen as particularly useful in that country). Other groups, such as the US-based National Wildlife Federation, the Nature Conservancy, and The Wilderness Society, and global groups like the World Wide Fund for Nature and Friends of the Earth, disseminate information, participate in public hearings, lobby, stage demonstrations, and may purchase land for preservation. Statewide nonprofit organizations such as the Wyoming Outdoor Council often collaborate with these national organizations and employ similar strategies. Smaller groups, including Wildlife Conservation International, conduct research on endangered species and ecosystems. More radical organizations, such as Greenpeace, Earth First!, and the Earth Liberation Front, have more directly opposed actions they regard as environmentally harmful. While Greenpeace is devoted to nonviolent confrontation as a means of bearing witness to environmental wrongs and bringing issues into the public realm for debate, the underground *Earth Liberation Front* engages in the clandestine destruction of property, the release of caged or penned animals, and other criminal acts. Such tactics are regarded as unusual within the movement, however.

On an international level, concern for the environment was the subject of a United Nations Conference on the Human Environment in Stockholm in 1972, attended by 114 nations. Out of this meeting developed UNEP (United Nations Environment Programme) and the follow-up United Nations Conference on Environment and Development in 1992. Other international organizations in support of environmental policies development include the Commission for Environmental Co-operation (as part of NAFTA), the European Environment Agency (EEA), and the Intergovernmental Panel on Climate Change (IPCC).

Environmental Protests

Climate activists blockade British Airports Authority's headquarters for day of action

Notable environmental protests and campaigns include:

- 2010 Xinfa aluminum plant protest
- Anti-WAAhnsinns Festival
- Camp for Climate Action
- Campaign against Climate Change
- Climate Rush
- Cofán people oil drilling protest (Ecuador)
- Earth Day
- Earth First!
- Earthlife Africa
- Global Day of Action
- Gurindji Strike
- Hands off our Forest

- Homes before Roads
- Kupa Piti Kungka Tjuta
- Love Canal protests
- March Against Monsanto
- Nevada Desert Experience
- Plane Stupid
- Plane Mad
- Qidong protest
- Save Manapouri Campaign
- Say Yes demonstrations
- Shifang protest
- Stop Climate Chaos

"March Against Monsanto", Vancouver, Canada, May 25, 2013

Assassinations

Every year, more than 100 environmental activists are murdered throughout the world.

In Popular Culture

- The popular media have been used to convey conservation messages in the U.S. For instance, the U.S. Forest Service created Smokey the Bear in 1944; he appeared in countless posters, radio and television programs, movies, press releases, and other guises to warn about forest fires. The comic strip *Mark Trail*, by environmentalist Ed Dodd, began in 1946; it still appears weekly in 175 newspapers. Another example is the children's animated show *Captain Planet and the Planeteers*, created by Ted Turner and Barbara Pyle in 1989 to inform children about environmental issues. The show aired for six seasons and 113 episodes, in 100 countries worldwide from 1990 to 1996.
- Miss Earth is one of the three largest international beauty pageants alongside Miss Universe and Miss World that promotes Environmental Awareness. The reigning titleholders dedicate their year to promote specific projects and often address issues concerning the environment and other global issues through school tours, tree planting activities, street campaigns, coastal clean ups, speaking engagements, shopping mall tours, media guesting, environmental fair, storytelling programs, eco-fashion shows, and other environmental activities. The Miss Earth winner is the spokesperson for the Miss Earth Foundation, the United Nations Environment Programme (UNEP) and other environ-

mental organizations. The Miss Earth Foundation also works with the environmental departments and ministries of participating countries, various private sectors and corporations, as well as Greenpeace and the World Wildlife Foundation (WWF).

An Alternative View

Many environmentalists believe that human interference with 'nature' should be restricted or minimised as a matter of urgency (for the sake of life, or the planet, or just for the benefit of the human species), whereas environmental skeptics and anti-environmentalists do not believe that there is such a need. One can also regard oneself as an environmentalist and believe that human 'interference' with 'nature' should be *increased*. Nevertheless, there is a risk that the shift from emotional environmentalism into the technical management of natural resources and hazards could decrease the touch of humans with nature, leading to less concern with environment preservation.

Localism (Politics)

Localism describes a range of political philosophies which prioritize the local. Generally, localism supports local production and consumption of goods, local control of government, and promotion of local history, local culture and local identity. Localism can be contrasted with regionalism and centralized government, with its opposite being found in the unitary state.

Localism can also refer to a systematic approach to organizing a national government so that local autonomy is retained rather than following the usual pattern of government and political power becoming centralized over time. This view of Localism is defined in the work *Localism, A Philosophy of Government*. This work is available in the third edition as an ebook, while an earlier hardcover edition, circulated underground without an ISBN, is rather rare.

On a conceptual level, there are important affinities between localism and deliberative democracy. This concerns mainly the democratic goal of engaging citizens in decisions that affect them.

History

Localists assert that throughout the world's history, most social and economic institutions have been scaled at the local level, as opposed to regional, interregional, or global (basically until the late 19th to the early 20th centuries). Only with imperialism and the industrial revolution did local scales become denigrated. Most proponents of localism position themselves as defending aspects of this earlier way of life; the phrase "relocalization" is often used in this sense.

In the 20th century, localism drew heavily on the writings of Leopold Kohr, E.F. Schumacher, Wendell Berry, and Kirkpatrick Sale, among others. More generally, localism draws on a wide range of movements and concerns and it proposes that by re-localizing democratic and economic relationships to the local level, social, economic and environmental problems will be more definable and solutions more easily created. They include anarchism, bioregionalism, environmentalism, the Greens, and more specific concerns about food, monetary policy and education. Political parties of all persuasions have also occasionally favored the devolution of power to local authorities. In this vein Alan Milburn, a Labour Party MP, has spoken of "making services more locally accountable, devolving more power to local communities and, in the process, forging a modern relationship between the state, citizens and services"

Beginning in the 1980s, a particularly visible strain of localism in the United States was a movement to buy locally produced products. This movement originated with organic farming and likely gained impetus because of growing dissatisfaction with organic certification and the failing economic model of industrial agriculture for small farmers. While the advocates of local consumption draw on protectionist arguments, they also appealed primarily to an environmental argument: that pollution caused by transporting goods was a major externality in a global economy, and one that "localvores" could greatly diminish. Also, environmental issues can be addressed when decision making power is held by those affected by the issues instead of power sources that do not understand the needs of local communities.

Political Philosophy

Localism as a philosophy is related to the principle of subsidiarity.

In the early 21st century, localists have frequently found themselves aligned with critics of globalisation. Variants of localism are prevalent within the Green movement. According to an article in *International Socialism*, localism of this sort seeks to "answer to the problems created by globalisation" with "calls to minimise international trade and to seek to establish economies based on 'local' self-sufficiency only."

Some localists believe that society should be organised politically along community lines, with each community being free to conduct its own business in whatever fashion its people see fit. The size of the communities is defined such that their members are both familiar and dependent on each other, a size something along the lines of a small town or village.

In reference to localism, Edward Goldsmith, former editor of *The Ecologist* magazine, claims: "The problems facing the world today can only be solved by restoring the functioning of those natural systems which once satisfied our needs, i.e. by fully exploiting those incomparable resources which are individual people, families, communities and ecosystems, which together make up the biosphere or real world"

Tip O'Neill, a longtime Speaker of the House in the US Congress, once famously declared that "All politics is local" He eventually wrote a book by that name: *All Politics Is Local: And Other Rules of the Game*.

Localism and Third World

Many localists are concerned with the problems of the development of the Third World. Many advocate that third world countries should aim to rely on their own goods and services to escape from what they see are the unfair trade relations with the developed world. George Monbiot claims this idea does not recognise the fact that, even if Third World countries often get a raw deal in trade relations, refusing to trade at all would be a significant blow, as the countries need the revenue generated by trade.

Some localists are also against immigration from poor countries to rich ones. One of the problems they claim results from such immigration is the drain on the intellectual resources of poor countries, so called brain drain. For example, in the past decade, Bulgaria is estimated to have lost more than 50,000 qualified scientists and skilled workers through emigration every year. About a fifth of them were highly educated specialists in chemistry, biology, medicine and physics.

International Relations

Some localists are against political intervention and peace keeping measures. They believe that communities should find solutions to their own problems and in their own time, in whatever fashion they decide. They believe that all societies are capable of achieving long term peace once given the opportunity to do so.

Localist Activism

Localism usually describes social measures or trends which emphasise or value local and small-scale phenomena. This is in contrast to large, all-encompassing frameworks for action or belief. Localism can therefore be contrasted with globalisation, and in some cases localist activism has parallels with opposition to corporate-led globalization. Localism can be geographical, but there are also transnational linkages. Localist movements are often organized in support of locally owned, independent businesses and nonprofit organizations. Although the focus of this aspect of localist activism is on "buy local," "support local food," and "bank local" campaigns, some organizations and businesses also combine the goal of increased local ownership with environmental sustainability and social fairness goals.

Examples of localism are:

- Support for local food networks, such as farmers' markets, community-supported agriculture, community gardens, farm-to-table programs, food cooperatives, and restaurants that serve local food. The slow food movement, using diverse, seasonal, natural food in reaction to multinational merchandising of food which is uniform, produced using industrial methods, and called fast food.
- Support for locally owned, independent businesses, including community banks and credit unions, such as the following organizations: American Independent Business Alliance, Business Alliance for Local Living Economies, the Institute for Local Self-Reliance, and Move Your Money. "The Benefits of Doing Business Locally" an essay by Jeff Milchen, American Independent Business Alliance co-founder, covers many arguments for local business ownership and patronage.
- Localism in media to support a diverse news media in the face of increasing corporate control. The US Federal Communications Commission is using this term when seeking input on its rules and states that "promoting localism is a key goal of the Commission's media ownership rules."
- Localism in government structures, which can include:
    - o Tertiary government where small community councils make relevant decisions, with some degree of independence from local or national government.

- Workers' councils, where the employees of a particular workplace discuss and negotiate with their employer, rather have this done by a national union which may be remote from local issues.
- Federalism and devolution.

- Postmodernism can be seen as a sort of cultural localism, where accepted cultural values may be ignored in favour of people creating their own criteria of value.
- Religion (Protestant):
    - Exclusive localism holds that there can't be more than one legitimate institutionally visible church at one given location, the variation of which varies but is usually held to be either a city or a neighbourhood.
    - Localism is more generally the congregationalist idea that each local church should be autonomous, only extended to reject any formal association of churches. It is specially relevant among Baptists, where localists reject the forming of Conventions.
- Religion (Churches of Christ):
    - The congregationalist idea of local autonomy is a cornerstone of restoration movement fellowships that identify as churches of Christ or Independent Christian Churches/Churches of Christ. Founders of the movement declared their independence from various denominations, seeking a fresh start to restore the New Testament church, and abandoning creeds. The names "Church of Christ," "Christian Church" and "Disciples of Christ" were adopted by the movement because they believed these terms to be biblical and not man made.
    - A converging of Christians across denominational lines in search of a return to a hypothesized original, "pre-denominational" Christianity. Participants in this movement sought to base doctrine and practice on the Bible alone, rather than recognizing the traditional councils and denominational hierarchies that had come to define Christianity since the 1st century AD. Members of the Churches of Christ believe that Jesus founded only one church, that the current divisions between Christians are not God's will, and that the only basis for restoring Christian unity is the Bible. They typically prefer to be known simply as "Christians", without any further religious or denominational identification. They see themselves as recreating the New Testament church established by Christ.
- Churches of Christ generally share these theological beliefs:
    - Refusal to hold to any formalized creeds or statements of faith, preferring instead a reliance on the Bible alone for doctrine and practice;
    - Autonomous, congregational church organization without denominational oversight;
    - Local governance by a plurality of male elders;

- One of the largest divisions within churches of Christ was due to controversy of foreign missionary work. Opponents of what they dubbed "Institutionalism" argued against it both as a drain on local congregations and as sinful if done in cooperation with other congregations. This belief extended to cooperative support of orphanages, homes, large-scale radio and TV programs and ministries.
- The Restoration Movement is so averse to association with other congregations that they renounce the term "protestant" distancing their churches from any association to any denomination; even one they would have to "protest" and evolve from.

- Localism as it applies to saving our globe economically and environmentally: Localism is essential in describing the magnitude of our climate crisis and economic concerns. Through several selfish actions, globally, we have created an entirely new planet; a planet that is struggling to maintain it's sustainability. The book, "Earth" by Bill McKibben, focusses on this topic. McKibben notes that we are at a point in which we need to drastically alter our attitude in order to provide the generations to come with a more bearable environment. This meaning, that our sights need to be directed towards merely just enduring, rather than growing. To do so, he suggests that we must return to, and rely on the notion of localism. As a result, communities will be able to rediscover what we may have lost sight of in lieu of the recent economic growth, that being the benefits of relying on our neighbors. An example of this is local farming; although it has shown to be more labor concentrated, in the long run it could prove to be more practical in an economic sense. McKibben believes that by supporting local farming, communities can essentially cut out the middlemen that are needed to transport, store, and package the items that we see in the supermarket and therefore save not only economically, but also conserve the limited energy resources that we do have. An article from Business and The Environment Journal in September 2011 stated that, "buying local makes consumers feel good," and "shorter trips from the field to the market mean fewer carbon emissions." Although this may depend on the product of interest, a decrease of carbon emission will lead to conserved energy which was McKibben's initial topic of concern in getting our world back to a more sustainable place. Fred Curtis from the Department of Economics at Drew University labels this broad notion as "eco-localism". He believes that "the road to environmental sustainability lies in the creation of local, self-reliant, community economies." Socially, physically, and financially, communities need to adapt and become more locally oriented in order for this process to play out favorably in the long run. One of the key points that Curtis acknowledges is that by producing goods and services based on local needs, communities can essentially produce higher quality products, in turn allowing the needs of consumers to be met, inevitably leading to a restructure of the local economy.

Ecocentrism

Ecocentrism is a term used in ecological political philosophy to denote a nature-centered, as opposed to human-centered (i.e. anthropocentric), system of values. The justification for ecocentrism usually consists in an ontological belief and subsequent ethical claim. The ontological belief denies that there are any existential divisions between human and non-human nature sufficient to claim that humans are either (a) the sole bearers of intrinsic value or (b) possess greater intrinsic value than non-human nature. Thus the subsequent ethical claim is for an equality of intrinsic value across human and non-human nature, or 'biospherical egalitarianism'. According to Stan Rowe:

> *The ecocentric argument is grounded in the belief that, compared to the undoubted importance of the human part, the whole ecosphere is even more significant and consequential: more inclusive, more complex, more integrated, more creative, more beautiful, more mysterious, and older than time. The "environment" that anthropocentrism misperceives as materials designed to be used exclusively by humans, to serve the needs of humanity, is in the profoundest sense humanity's source and support: its ingenious, inventive life-giving matrix. Ecocentrism goes beyond biocentrism with its fixation on organisms, for in the ecocentric view people are inseparable from the inorganic/organic nature that encapsulates them. They are particles and waves, body and spirit, in the context of Earth's ambient energy.*

and:

> *To switch Western culture from its present track to a saving ecopolitical route means finding a new and compelling belief-system to redirect our way-of-living. It must be a vital outgrowth from our science-based culture. It seems to me that the only promising universal belief-system is ecocentrism, defined as a value-shift from* Homo sapiens *to planet earth. A scientific rationale backs the value-shift. All organisms are evolved from Earth, sustained by Earth. Thus Earth, not organism, is the metaphor for Life. Earth not humanity is the Life-center, the creativity-center. Earth is the whole of which we are subservient parts. Such a fundamental philosophy gives ecological awareness and sensitivity an enfolding, material focus.*
>
> *Ecocentrism is not an argument that all organisms have equivalent value. It is not an anti-human argument nor a put-down of those seeking social justice. It does not deny that myriad important homocentric problems exist. But it stands aside from these smaller, short-term issues in order to consider Ecological Reality. Reflecting on the ecological status of all organisms, it comprehends the Ecosphere as a Being that transcends in importance any one single species, even the self-named sapient one.* (1994)

Origin of Term

The ecocentric ethic was conceived by Aldo Leopold and recognizes that all species, including humans, are the product of a long evolutionary process and are inter-related in their life processes. The writings of Aldo Leopold and his idea of the land ethic and good environmental management are a key element to this philosophy. Ecocentrism focuses on the biotic community as a whole and strives to maintain ecosystem composition and ecological processes. The term also finds expression in the first principle of the deep ecology movement, as formulated by Arne Næss and George Sessions in 1984 which points out that anthropocentrism, which considers humans as the center of the universe and the pinnacle of all creation, is a difficult opponent for ecocentrism.

Background

Environmental thought and the various branches of the environmental movement are often classified into two intellectual camps: those that are considered anthropocentric, or "human-centred," in orientation and those considered biocentric, or "life-centred". This division has been described in other terminology as "shallow" ecology versus "deep" ecology and as "technocentrism" versus "ecocentrism". Ecocentrism can be seen as one stream of thought within environmentalism, the political and ethical movement that seeks to protect and improve the quality of the natural environment through changes to environmentally harmful human activities by adopting environmentally benign forms of political, economic, and social organization and through a reassessment of humanity's relationship with nature. In various ways, environmentalism claims that non-human organisms and the natural environment as a whole deserve consideration when appraising the morality of political, economic, and social policies.

Relationship to Other Similar Philosophies

Anthropocentrism

Ecocentrism is taken by its proponents to constitute a radical challenge to long-standing and deeply rooted anthropocentric attitudes in Western culture, science, and politics. Anthropocentrism is alleged to leave the case for the protection of non-human nature subject to the demands of human utility, and thus never more than contingent on the demands of human welfare. An ecocentric ethic, by contrast, is believed to be necessary in order to develop a non-contingent basis for protecting the natural world. Critics of ecocentrism have argued that it opens the doors to an anti-humanist morality that risks sacrificing human well-being for the sake of an ill-defined 'greater good'. Deep ecologist Arne Naess has identified anthropocentrism as a root cause of the ecological crisis, human overpopulation, and the extinctions of many non-human species. Others point to the gradual historical realization that humans are not the centre of all things, that "A few hundred years ago, with some reluctance, Western people admitted that the planets, Sun and stars did not circle around their abode. In short, our thoughts and concepts though irreducibly anthropomorphic need not be anthropocentric."

Industrocentrism

Industrocentrism is an ideology that goes hand in hand with today's industrial neoliberal capitalist agenda. It sees all things on earth as resources to be utilized by humans or to be commodified. This view is the opposite of anthropocentrism and ecocentrism. It negatively affects humans, nonhumans, and the environment in the long run in that it only focuses on short term economic gratification.

Technocentrism

Ecocentrism is also contrasted with technocentrism (meaning values centred on technology) as two opposing perspectives on attitudes towards human technology and its ability to affect, control and even protect the environment. Ecocentrics, including "deep green" ecologists, see themselves as being subject to nature, rather than in control of it. They lack faith in modern technology and the bureaucracy attached to it. Ecocentrics will argue that the natural world should be respected for its processes and products, and that low impact technology and self-reliance is more desirable than technological control of nature. Technocentrics, including imperialists, have absolute faith in technology and industry and firmly believe that humans have control over nature. Although technocentrics may accept that environmental problems do exist, they do not see them as problems to be solved by a reduction in industry. Rather, environmental problems are seen as problems to be solved using science. Indeed, technocentrics see that the way forward for developed and developing countries and the solutions to our environmental problems today lie in scientific and technological advancement.

Biocentrism

The distinction between biocentrism and ecocentrism is ill-defined. Ecocentrism recognizes Earth's interactive living and non-living systems rather than just the Earth's organisms (biocentrism) as central in importance. The term has been used by those advocating "left biocentrism", combining deep ecology with an "anti-industrial and anti-capitalist" position (David Orton et al.).

References

- Aboul-Enein, H. Yousuf; Zuhur, Sherifa (2004), Islamic Rulings on Warfare, Strategic Studies Institute, US Army War College, Diane Publishing Co., Darby PA, p. 22, ISBN 9781584871774
- Thomas P. Govan, "Agrarian and Agrarianism: A Study in the Use and Abuse of Words," Journal of Southern History, Vol. 30#1 (Feb., 1964), pp. 35–47 in JSTOR
- Cat Lincoln (Spring 2009). "Light, Dark and Bright Green Environmentalism". Green Daily. Archived from the original on 25 April 2009. Retrieved 2009-11-02
- Thomas R. DeGregori (2002). Bountiful Harvest: Technology, Food Safety, and the Environment. Cato Institute. p. 153. ISBN 1-930865-31-7
- Gari, L. (November 2002), "Arabic Treatises on Environmental Pollution up to the End of the Thirteenth Century", Environment and History, 8 (4): 475–488, doi:10.3197/096734002129342747
- David Urbinato (Summer 1994). "London's Historic 'Pea-Soupers'". United States Environmental Protection Agency. Archived from the original on 2 October 2006. Retrieved 2006-08-02
- Huesemann, Michael H., and Joyce A. Huesemann (2011). Technofix: Why Technology Won't Save Us or the Environment, New Society Publishers, Gabriola Island, British Columbia, Canada, ISBN 0865717044, 464 pp
- McCallum, M.L. & G.W. Bury (2013). "Google search patterns suggest declining interest in the environment". Biodiversity and Conservation. 22 (6): 1355–1367. doi:10.1007/s10531-013-0476-6
- Hess, David J. (2009). Localist Movements in a Global Economy: Sustainability, Justice, and Urban Development in the United States. Cambridge, MA: MIT Press. ISBN 9780262512329
- Tom Brooking, "'Busting Up' the Greatest Estate of All: Liberal Maori Land Policy, 1891–1911," New Zealand Journal of History (1992) 26#1 pp. 78–98
- Lindenmeyer, D. & Burgman, M. 2005. Practical conservation biology. CSIRO Publishing, Collingwood, Australia. ISBN 0-643-09089-4
- "Vasconcelos, Vitor Vieira "The Environment Professional and the Touch with Nature." Qualit@s, v 1, p. 1-10, 2011". Retrieved 2013-05-15
- Shoreman-Ouimet, Eleanor; Kopnina, Helen (2016). Culture and Conservation: Beyond Anthropocentrism. Oxon: Routledge. pp. 8–9. ISBN 9781315858630
- Ercan, S.A.; Hendriks, C.H. (2013). "The democratic challenges and potential of localism: Insights from deliberative democracy". Policy Studies. 31 (4): 422–440. doi:10.1080/01442872.2013.822701

4

Values and Principles of Green Politics

The main values and principles of green politics are pacifism, peace movement, social justice and social equality. Pacifism is the condemnation of war and violence whereas peace movement is a movement which focuses on ending war and promoting peace. This chapter elucidates the values and principles of green politics.

Pacifism

A peace sign, which is widely associated with pacifism

Pacifism is opposition to war, militarism, or violence. The word *pacifism* was coined by the French peace campaigner Émile Arnaud (1864–1921) and adopted by other peace activists at the tenth Universal Peace Congress in Glasgow in 1901. A related term is *ahimsa* (to do no harm), which is a core philosophy in Buddhism, Jainism, and Hinduism. While modern connotations are recent, having been explicated since the 19th century, ancient references abound.

In modern times, interest was revived by Leo Tolstoy in his late works, particularly in *The Kingdom of God Is Within You*. Mohandas Gandhi (1869–1948) propounded the practice of steadfast nonviolent opposition which he called "satyagraha", instrumental in its role in the Indian Independence Movement. Its effectiveness served as inspiration to Martin Luther King Jr., James Lawson, James Bevel, Thich Nhat Hanh and many others in the Civil Rights Movement. Pacifism was widely associated with the much publicized image of Tiananmen Square Protests of 1989 with the "Tank Man", where one protester stood in nonviolent opposition to a column of tanks.

Definition

Pacifism covers a spectrum of views, including the belief that international disputes can and should be peacefully resolved, calls for the abolition of the institutions of the military and war, opposition to any organization of society through governmental force (anarchist or libertarian pacifism), rejection of the use of physical violence to obtain political, economic or social goals, the obliteration of force, and opposition to violence under any circumstance, even defence of self and others. Historians of pacifism Peter Brock and Thomas Paul Socknat define pacifism "in the sense generally accepted in English-speaking areas" as "an unconditional rejection of all forms of warfare". Philosopher Jenny Teichman defines the main form of pacifism as "anti-warism", the rejection of all forms of warfare. Teichman's beliefs have been summarized by Brian Orend as "... A pacifist rejects war and believes there are no moral grounds which can justify resorting to war. War, for the pacifist, is always wrong." In a sense the philosophy is based on the idea that the ends do not justify the means.

Moral Considerations

Anti-war activist arrested in San Francisco during the March 2003 protests against the war in Iraq

Pacifism may be based on moral principles (a deontological view) or pragmatism (a consequentialist view). Principled pacifism holds that at some point along the spectrum from war to interpersonal physical violence, such violence becomes morally wrong. Pragmatic pacifism holds that the costs of war and interpersonal violence are so substantial that better ways of resolving disputes must be found, Pacifists generally reject theories of Just War.

Nonviolence

Some pacifists follow principles of nonviolence, believing that nonviolent action is morally superior and/or most effective. Some however, support physical violence for emergency defence of self or others. Others support destruction of property in such emergencies or for conducting symbolic acts of resistance like pouring red paint to represent blood on the outside of military recruiting offices or entering air force bases and hammering on military aircraft.

By no means is all nonviolent resistance (sometimes also called civil resistance) based on a fundamental rejection of all violence in all circumstances. Many leaders and participants in such movements, while recognizing the importance of using non-violent methods in particular circumstances, have not been absolute pacifists. Sometimes, as with the civil rights movement's march from Selma to Montgomery in 1965, they have called for armed protection. The interconnections between civil resistance and factors of force are numerous and complex.

Absolute Pacifism

An absolute pacifist is generally described by the British Broadcasting Corporation as one who believes that human life is so valuable, that a human should never be killed and war should never be conducted, even in self-defense. The principle is described as difficult to abide by consistently, due to violence not being available as a tool to aid a person who is being harmed or killed. It is further claimed that such a pacifist could logically argue that violence leads to more undesirable results than non-violence.

Police Actions and National Liberation

Although all pacifists are opposed to war between nation states, there have been occasions where pacifists have supported military conflict in the case of civil war or revolution. For instance, during the American Civil War, both the American Peace Society and some former members of the Non-Resistance Society supported the Union's military campaign, arguing they were carrying out a "police action" against the Confederacy, whose act of Secession they regarded as criminal. Following the outbreak of the Spanish Civil War, French pacifist René Gérin (1892–1957) urged support for the Spanish Republic. Gérin argued that the Spanish Nationalists were "comparable to an individual enemy" and the Republic's war effort was equivalent to the action of a domestic police force suppressing crime.

In the 1960s, some pacifists associated with the New Left supported wars of national liberation and supported groups such as the Viet Cong and the Algerian FLN, arguing peaceful attempts to liberate such nations were no longer viable, and war was thus the only option.

Early Traditions of Pacifism

Vereshchagin's painting *Apotheosis of War* (1871) came to be admired as one of the earliest artistic expressions of pacifism

Advocacy of pacifism can be found far back in history and literature.

China

During the Warring States period, the pacifist Mohist School opposed aggressive war between the feudal states. They took this belief into action by using their famed defensive strategies to defend smaller states from invasion from larger states, hoping to dissuade feudal lords from costly warfare. The Seven Military Classics of ancient China view warfare negatively, and as a last resort. For example, the *Three Strategies of Huang Shigong* says: "As for the military, it is not an auspicious instrument; it is the way of heaven to despise it", and the *Wei Liaozi* writes: "As for the military, it is an inauspicious instrument; as for conflict and contention, it runs counter to virtue".

The Taoist scripture "*Classic of Great Peace* (*Taiping jing*)" foretells "the coming Age of Great Peace (*taiping*)." The *Taiping Jing* advocates "a world full of peace".

Lemba

The Lemba religion of southern French Congo, along with its symbolic herb, is named for pacifism : "*lemba, lemba*" (peace, peace), describes the action of the plant *lemba-lemba* (*Brillantaisia patula T. Anders*). Likewise in Cabinda, "*Lemba* is the spirit of peace, as its name indicates."

Moriori

Moriori tree carving found in the Chatham Islands.

The Moriori, of the Chatham Islands, practiced pacifism by order of their ancestor Nunuku-whenua. This enabled the Moriori to preserve what limited resources they had in their harsh climate, avoiding waste through warfare. In turn, this led to their almost complete annihilation in 1835 by invading Ngāti Mutunga and Ngāti Tama Māori from the Taranaki region of the North Island of New Zealand. The invading Māori killed, enslaved and cannibalised the Moriori. A Moriori survivor recalled : "[The Maori] commenced to kill us like sheep ... [We] were terrified, fled to the bush, concealed ourselves in holes underground, and in any place to escape our enemies. It was of no avail; we were discovered and killed - men, women and children indiscriminately."

Greece

In Ancient Greece, pacifism seems not to have existed except as a broad moral guideline against violence between individuals. No philosophical program of rejecting violence between states, or rejecting all forms of violence, seems to have existed. Aristophanes, in his play Lysistrata, creates the scenario of an Athenian woman's anti-war sex strike during the Peloponnesian War of 431–404 BC, and the play has gained an international reputation for its anti-war message. Nevertheless, it is both fictional and comical, and though it offers a pragmatic opposition to the destructiveness of war, its message seems to stem from frustration with the existing conflict (then in its twentieth year) rather than from a philosophical position against violence or war. Equally fictional is the nonviolent protest of Hegetorides of Thasos. Euripides also expressed strong anti-war ideas in his work, especially *The Trojan Women*.

Roman Empire

Several Roman writers rejected the militarism of Roman society and gave voice to anti-war sentiments, including Propertius, Tibullus and Ovid. The Stoic Seneca the Younger criticised warfare in his book *Naturales quaestiones* (circa 65 AD).

Maximilian of Tebessa was a Christian conscientious objector. He was killed for refusing to be conscripted.

Christianity

Throughout history many have understood Jesus of Nazareth to have been a pacifist, drawing on his Sermon on the Mount. In the sermon Jesus stated that one should "not resist an evildoer" and promoted his turn the other cheek philosophy. "If anyone strikes you on the right cheek, turn the other also; and if anyone wants to sue you and take your coat, give your cloak as well ... Love your enemies, do good to those who hate you, bless those who curse you, pray for those who abuse you." The New Testament story is of Jesus, besides preaching these words, surrendering himself freely to an enemy intent on having him killed and proscribing his followers from defending him.

There are those, however, who deny that Jesus was a pacifist and state that Jesus never said not to fight, citing examples from the New Testament. One such instance portrays an angry Jesus driving dishonest market traders from the temple. A frequently quoted passage is Luke 22:36: "He said to them, 'But now, the one who has a purse must take it, and likewise a bag. And the one who has no sword must sell his cloak and buy one.'" Pacifists have typically explained that verse as Jesus fulfilling prophecy, since in the next verse, Jesus continues to say: "It is written: 'And he was numbered with the transgressors'; and I tell you that this must be fulfilled in me. Yes, what is written about me is reaching its fulfillment." Others have interpreted the non-pacifist statements in the New Testament to be related to self-defense or to be metaphorical and state that on no occasion did Jesus shed blood or urge others to shed blood.

Cathars

Known in the Balkans as Bogomils and in northern Italy and southern France as Cathars, they were pacifists totally dedicated to nonviolence. The Cathars were actually branded heretics, persecuted, and eventually annihilated by the Catholic Church through the Albigensian Crusade and the Inquisition that followed. "These heretics are worse than the saracens" exclaimed Pope Innocent III, and on March 10, 1208, after the murder of the papal legate Pierre de Castelnau, probably by Raymond VI, Count of Toulouse, the pope took full advantage of it and proclaimed a crusade against a sect in southern France.

Modern History

Penn's Treaty with the Indians. This treaty was never violated.

Beginning in the 16th century, the Protestant Reformation gave rise to a variety of new Christian sects, including the historic peace churches. Foremost among them were the Religious Society of Friends (Quakers), Amish, Mennonites, and Church of the Brethren. The humanist writer Desiderius Erasmus was one of the most outspoken pacifists of the Renaissance, arguing strongly against warfare in his essays *The Praise of Folly* (1509) and *The Complaint of Peace* (1517).

The Quakers were prominent advocates of pacifism, who as early as 1660 had repudiated violence in all forms and adhered to a strictly pacifist interpretation of Christianity. They stated their beliefs in a declaration to King Charles II:

"We utterly deny all outward wars and strife, and fightings with outward weapons, for any end, or under any pretense whatever; this is our testimony to the whole world. The Spirit of Christ ... which leads us into all truth, will never move us to fight and war against any man with outward weapons, neither for the kingdom of Christ, nor for the kingdoms of this world.

Throughout the many 18th century wars in which Britain participated, the Quakers maintained a principled commitment not to serve in the army and militia or even to pay the alternative £10 fine.

The English Quaker William Penn, who founded the Province of Pennsylvania, employed an anti-militarist public policy. Unlike residents of many of the colonies, Quakers chose to trade peacefully with the Indians, including for land. The colonial province was, for the 75 years from 1681 to 1756, essentially unarmed and experienced little or no warfare in that period.

From the 16th to the 18th centuries, a number of thinkers devised plans for an international organisation that would promote peace, and reduce or even eliminate the occurrence of war. These included the French politician Duc de Sully, the philosophers Émeric Crucé and the Abbe de Saint-Pierre, and the English Quakers William Penn and John Bellers.

Pacifist ideals emerged from two strands of thought that coalesced at the end of the 18th century. One, rooted in the secular Enlightenment, promoted peace as the rational antidote to the world's ills, while the other was a part of the evangelical religious revival that had played an important part in the campaign for the abolition of slavery. Representatives of the former, included Jean-Jacques

Rousseau, in *Extrait du Projet de Paix Perpetuelle de Monsieur l'Abbe Saint-Pierre* (1756), Immanuel Kant, in his *Thoughts on Perpetual Peace* and Jeremy Bentham who proposed the formation of a peace association in 1789. Representative of the latter, was William Wilberforce who thought that strict limits should be imposed on British involvement in the French Revolutionary War based on Christian ideals of peace and brotherhood. Bohemian Bernard Bolzano (1781–1848) taught about the social waste of militarism and the needlessness of war. He urged a total reform of the educational, social, and economic systems that would direct the nation's interests toward peace rather than toward armed conflict between nations.

Peace Movements

During the period of the Napoleonic Wars, although no formal peace movement was established until the end of hostilities, a significant peace movement animated by universalist ideals did emerge, due to the perception of Britain fighting in a reactionary role and the increasingly visible impact of the war on the welfare of the nation in the form of higher taxation levels and high casualty rates. Sixteen peace petitions to Parliament were signed by members of the public, anti-war and anti-Pitt demonstrations convened and peace literature was widely published and disseminated.

"Peace". Caricature of Henry Richard, a prominent advocate of pacifism in the mid-19th century

The first peace movements appeared in 1815–16. In the United States the first such movement was the New York Peace Society, founded in 1815 by the theologian David Low Dodge, and the Massachusetts Peace Society. It became an active organization, holding regular weekly meetings, and producing literature which was spread as far as Gibraltar and Malta, describing the horrors of war and advocating pacificism on Christian grounds. The London Peace Society (also known as the Society for the Promotion of Permanent and Universal Peace) was formed in 1816 to promote permanent and universal peace by the philanthropist William Allen. In the 1840s, British women formed "Olive Leaf Circles", groups of around 15 to 20 women, to discuss and promote pacifist ideas.

The peace movement began to grow in influence by the mid-nineteenth century. The London Peace Society, under the initiative of American consul to Birmingham Elihu Burritt and the reverend Henry Richard, convened the first International Peace Congress in London in 1843. The congress decided on two aims: the ideal of peaceable arbitration in the affairs of nations and the creation of an international institution to achieve that. Richard became the secretary of the Peace Society in 1850 on a full-time basis, a position which he would keep for the next 40 years, earning himself a reputation as the 'Apostle of Peace'. He helped secure one of the earliest victories for the peace movement by securing a commitment from the Great Powers in the Treaty of Paris (1856) at the end of the Crimean War, in favour of arbitration. On the European continent, wracked by social upheaval, the first peace congress was held in Brussels in 1848 followed by Paris a year later.

After experiencing a recession in support due to the resurgence of militarism during the American Civil War and Crimean War, the movement began to spread across Europe and began to infiltrate the new working class socialist movements. In 1870, Randal Cremer formed the Workman's Peace Association in London. Cremer, alongside the French economist Frédéric Passy was also the founding father of the first international organisation for the arbitration of conflicts in 1889, the Inter-Parliamentary Union. The National Peace Council was founded in after the 17th Universal Peace Congress in London (July August 1908).

An important thinker who contributed to pacifist ideology was Russian writer Leo Tolstoy. In one of his latter works, *The Kingdom of God is Within You*, Tolstoy provides a detailed history, account and defense of pacifism. Tolstoy's work inspired a movement named after him advocating pacifism to arise in Russia and elsewhere. The book was a major early influence on Mohandas K. Gandhi (1869–1948), and the two engaged in regular correspondence while Gandhi was active in South Africa.

Bertha von Suttner, the first woman to be a Nobel Peace Prize laureate, became a leading figure in the peace movement with the publication of her novel, *Die Waffen nieder!* ("Lay Down Your Arms!") in 1889 and founded an Austrian pacifist organization in 1891.

Non-violent Resistance

"Leading Citizens want War and declare War; Citizens Who are Led fight the War" 1910 cartoon

In New Zealand, during the latter half of the 19th century British colonists used many tactics to confiscate land from the indigenous Māori, including warfare. In the 1870s and 1880s, Parihaka, then reputed to be the largest Māori village in New Zealand, became the centre of a major campaign of non-violent resistance to European occupation of confiscated land in the area. One Māori leader, Te Whiti-o-Rongomai, inspired warriors to stand up for their rights without using weapons, which had led to defeat in the past. In 1881 he convinced 2000 Maori to welcome battle-hardened British soldiers into their village and even offered food and drink. He allowed himself and his people to be arrested without resistance for opposing land confiscation. He is remembered as a great leader because the "passive resistance" his practice prevented British massacres and even protected far more land than violent resistance.

Mohandas K. Gandhi was a major political and spiritual leader of India, instrumental in the Indian independence movement. The Nobel prize winning great poet Rabindranath Tagore, who was also an Indian, gave him the honorific "Mahatma", usually translated "Great Soul." He was the pioneer of a brand of nonviolence (or *ahimsa*) which he called *satyagraha*—translated literally as "truth force". This was the resistance of tyranny through civil disobedience that was not only nonviolent but also sought to change the heart of the opponent. He contrasted this with *duragraha*, "resistant force," which sought only to change behaviour with stubborn protest.

During his 30 years of work (1917–1947) for the independence of his country from the British Raj, Gandhi led dozens of nonviolent campaigns, spent over seven years in prison, and fasted nearly to the death on several occasions to obtain British compliance with a demand or to stop inter-communal violence. His efforts helped lead India to independence in 1947, and inspired movements for civil rights and freedom worldwide.

World War I

The Deserter (1916) by Boardman Robinson

Peace movements became active in the Western world after 1900, often focusing on Treaties that would settle disputes through arbitration, and efforts to support the Hague conventions.

The sudden outbreak of the First World War in July 1914 dismayed the peace movement. Socialist parties in every industrial nation had committed themselves to antiwar policies, but when the war. In fact took place all of them, except in Russia and the United States, supported their own government. There were highly publicized dissidents, some of whom went to jail for opposing the fraft laws, like Eugene Debs in the U.S. In Britain, the prominent activist Stephen Henry Hobhouse went to prison for refusing military service, citing his convictions as an "International Socialist and a Christian" Many socialist groups and movements were antimilitarist, arguing that war by its nature was a type of governmental coercion of the working class for the benefit of capitalist elites. The French socialist pacifist leader Jean Jaurès was assassinated by a nationalist fanatic on July 31, 1914. The national parties in the Second International increasingly supported their respective nations in war and the International was dissolved in 1916.

In 1915 the League of Nations Society was formed by British liberal leaders to promote a strong international organisation that could enforce the peaceful resolution of conflict. Later that year the League to Enforce Peace was established in America to promote similar goals. Hamilton Holt published an editorial in his New York City weekly magazine the *Independent* called "The Way to Disarm: A Practical Proposal" on September 28, 1914. It called for an international organization to agree upon the arbitration of disputes and to guarantee the territorial integrity of its members by maintaining military forces sufficient to defeat those of any non-member. The ensuing debate among prominent internationalists modified Holt's plan to align it more closely with proposals offered in Great Britain by Viscount James Bryce, a former ambassador from the U.K. to the U.S. These and other initiatives were pivotal in the change in attitudes that gave birth to the League of Nations after the war.

A World War I-era female peace protester

Some of the many groups that protested against the war, as well as the traditional peace churches, were the Woman's Peace Party (which was organized in 1915 and led by noted reformer Jane Addams), the International Committee of Women for Permanent Peace (ICWPP) (also organized in 1915), the American Union Against Militarism, the Fellowship of Reconciliation, and the American Friends Service Committee. Jeannette Rankin, the first woman elected to Congress, was another fierce advocate of pacifism, the only person to vote no to America's entrance into both World Wars.

Between the two World Wars

The immense loss of life during the war, for what became regarded as futile reasons, caused a sea-change in public attitudes to militarism. Organisations formed in this period included the War Resisters' International the Women's International League for Peace and Freedom, the No More War Movement and the Peace Pledge Union (PPU). The League of Nations also convened several disarmament conferences in the inter-war period such as the Geneva Conference.

Pacifism and revulsion with war were very popular sentiments in 1920s Britain. A stream of novels and poems on the theme of the futility of war and the slaughter of the youth by old fools were published, including, Death of a Hero by Richard Aldington, Erich Remarque's translated All Quiet on the Western Front and Beverley Nichols's expose *Cry Havoc*. A debate at the University of Oxford in 1933 on the motion 'one must fight for King and country' captured the changed mood when the motion was resoundingly defeated. Dick Sheppard established the Peace Pledge Union in 1934 totally renouncing war and aggression. The idea of collective security was also popular; instead of outright pacifism the public generally exhibited a determination to stand up to aggression, but preferably with the use of economic sanctions and multilateral negotiations.

Refugees from the Spanish Civil War at the War Resisters' International children's refuge in the French Pyrenees

The British Labour Party had a strong pacifist wing in the early 1930s and between 1931 and 1935 was led by George Lansbury, a Christian pacifist who later chaired the No More War Movement and was president of the PPU. The 1933 annual conference resolved unanimously to "pledge itself to take no part in war". "Labour's official position, however, although based on the aspiration towards a world socialist commonwealth and the outlawing of war, did not imply a renunciation of force under all circumstances, but rather support for the ill-defined concept of 'collective security' under the League of Nations. At the same time, on the party's left, Stafford Cripps's small but vocal Socialist League opposed the official policy, on the non-pacifist ground that the League of Nations was 'nothing but the tool of the satiated imperialist powers'." Lansbury was eventually persuaded to resign as Labour leader by the non-pacifist wing of the party and was replaced by Clement Attlee. As the threat from Nazi Germany increased in the 1930s, the Labour Party abandoned its pacifist position and supported re-armament, largely due to the efforts of Ernest Bevin and Hugh Dalton who by 1937 had also persuaded the party to oppose Neville Chamberlain's policy of appeasement.

The League of Nations attempted to play its role of ensuring world peace in the 1920s and 30s, although with the increasingly revisionist and aggressive behaviour of Nazi Germany, Fascist Italy and Imperial Japan, it ultimately failed to maintain such a world order. Economic sanctions were used against states that committed aggression, such as Italy when it invaded Abyssinia, but there was no will on the part of the principal League powers, Britain and France, to subordinate their interests to a multilateral process or to disarm at all themselves.

The Spanish Civil War proved a major test for international pacifism, and the work of pacifist organisations (such as War Resisters' International and the Fellowship of Reconciliation) and individuals (such as José Brocca and Amparo Poch) in that arena has until recently been ignored or forgotten by historians, overshadowed by the memory of the International Brigades and other militaristic interventions. Shortly after the war ended, Simone Weil, despite having volunteered for service on the republican side, went on to publish *The Iliad or the Poem of Force*, a work that has been described as a pacifist manifesto. In response to the threat of fascism, some pacifist thinkers, such as Richard B. Gregg, devised plans for a campaign of nonviolent resistance in the event of a fascist invasion or takeover.

World War II

With the start of World War II, pacifist and anti-war sentiment declined in nations affected by war. Even the communist-controlled American Peace Mobilization reversed its anti-war activism once Germany invaded the Soviet Union in 1941. After the Japanese attack on Pearl Harbor, mainstream isolationist groups like the America First Committee, declined, but many smaller religious and socialist groups continued their opposition to war. Bertrand Russell argued that the necessity of defeating Adolf Hitler and the Nazis was a unique circumstance where war was not the worst of the possible evils; he called his position *relative pacifism*. Shortly before the outbreak of war, British writers such as E. M. Forster, Leonard Woolf, David Garnett and Storm Jameson all rejected their earlier pacifism and endorsed military action against Nazism. Similarly Albert Einstein wrote: "I loathe all armies and any kind of violence; yet I'm firmly convinced that at present these hateful weapons offer the only effective protection." The British pacifists Reginald Sorensen and C. J. Cadoux, while bitterly disappointed by the outbreak of war, nevertheless urged their fellow pacifists "not to obstruct the war effort".

The French pacifists André and Magda Trocmé helped conceal hundreds of Jews fleeing the Nazis in the village of Le Chambon-sur-Lignon. After the war, the Trocmés were declared Righteous Among the Nations.

Pacifists in the Third Reich were dealt with harshly; German pacifist Carl von Ossietzky, and Olaf Kullmann, a Norwegian pacifist active during the Nazi occupation, were both imprisoned in concentration camps and died as a result of their mistreatment there. Austrian farmer Franz Jägerstätter was executed in 1943 for refusing to serve in the Wehrmacht.

After the end of the war, it was discovered that "The Black Book" or *Sonderfahndungsliste G.B.* list of Britons to be arrested in the event of a Nazi invasion of the UK included three active pacifists; Vera Brittain, Sybil Thorndike and Aldous Huxley (who had left the country).

There were conscientious objectors and war tax resisters in both World War I and World War II. The United States government allowed sincere objectors to serve in noncombatant military roles. However, those draft resisters who refused any cooperation with the war effort often spent much of each war in federal prisons. During World War II, pacifist leaders like Dorothy Day and Ammon Hennacy of the Catholic Worker Movement urged young Americans not to enlist in military service.

Later Twentieth Century

Martin Luther King, Jr (1929–68), a Baptist minister, led the Civil Rights Movement which successfully used Gandhian nonviolent resistance to repeal laws enforcing racial segregation and work for integration of schools, businesses and government. In 1957, his wife Coretta Scott King, Albert Schweitzer, Benjamin Spock, and others formed the Committee for a Sane Nuclear Policy (now Peace Action) to resist the nuclear arms race. In 1958 British activists formed the Campaign for Nuclear Disarmament with Bertrand Russell as its president.

A demonstrator offers a flower to military police at an anti-Vietnam War protest, 1967

In 1960, Thich Nhat Hanh came to the US to study comparative religion at Princeton University and subsequently was appointed lecturer in Buddhism at Columbia University. Thich Nhat Hanh had written a letter to Martin Luther King in 1965 entitled "Searching for the Enemy of Man" and during his 1966 stay in the US met with King and urged him to publicly denounce the Vietnam War. King gave his famous speech at the Riverside Church in New York City in 1967, his first to publicly question the U.S. involvement in Vietnam.

Other examples from this period include the 1986 People Power Revolution in the Philippines led by Cory Aquino, and the 1989 Tiananmen Square Protests which included the broadly publicized "Tank Man" incident.

On December 1, 1948, President José Figueres Ferrer of Costa Rica abolished the Costa Rican military. In 1949, the abolition of the military was introduced in Article 12 of the Costa Rican constitution. The budget previously dedicated to the military is now dedicated to providing health care services and education.

Religious Attitudes

Ahmadiyya

According to the Ahmadiyya understanding of Islam, pacifism is a strong current, and jihad is one's personal inner struggle and should not be used violently for political motives. Violence is the last option only to be used to protect religion and one's own life in extreme situations of persecution. Mirza Ghulam Ahmad, the founder of the Ahmadiyya Muslim Community, said that in contrary to the current views, Islam *does not allow the use of sword in religion, except in the case of defensive wars, wars waged to punish a tyrant, or those meant to uphold freedom.*

Ahmadiyya claims its objective to be the peaceful propagation of Islam with special emphasis on spreading the true message of Islam by the pen. Ahmadis point out that as per prophecy, who they believe was the promised messiah, Mirza Ghulam Ahmad, rendered the concept of violent jihad unnecessary in modern times. They believe that the answer of hate should be given by love.

Bahá'í Faith

Bahá'u'lláh, the founder of the Bahá'í Faith abolished holy war and emphasized its abolition as a central teaching of his faith. However, the Bahá'í Faith does not have an absolute pacifistic position. For example, Bahá'ís are advised to do social service instead of active army service, but when this is not possible because of obligations in certain countries, the Bahá'í law of *loyalty to one's government* is preferred and the individual should perform the army service. Shoghi Effendi, the head of the Bahá'í Faith in the first half of the 20th century, noted that in the Bahá'í view, absolute pacifists are anti-social and exalt the individual over society which could lead to anarchy; instead he noted that the Bahá'í conception of social life follows a moderate view where the individual is not suppressed or exalted.

On the level of society, Bahá'u'lláh promotes the principle of collective security, which does not abolish the use of force, but prescribes "a system in which Force is made the servant of Justice." The idea of collective security from the Bahá'í teachings states that if a government violates a fundamental norm of international law or provision of a future world constitution which Bahá'ís believe will be established by all nations, then the other governments should step in.

Buddhism

Buddhism is generally seen as among the least violent religious traditions, and Ahimsa (do no harm), is a primary virtue in Buddhism.

Aung San Suu Kyi is a Buddhist nonviolent pro-democracy activist and leader of the National League for Democracy in Myanmar (Burma), who became State Counsellor (similar to prime minister) of Myanmar in April 2016. A devout Buddhist, Suu Kyi won the Rafto Prize and the Sakharov Prize for Freedom of Thought in 1990 and in 1991 was awarded the Nobel Peace Prize for her peaceful and non-violent struggle under a repressive military dictatorship. One of her best known speeches is the "Freedom From Fear" speech, which begins, "It is not power that corrupts but fear. Fear of losing power corrupts those who wield it and fear of the scourge of power corrupts those who are subject to it."

Christianity

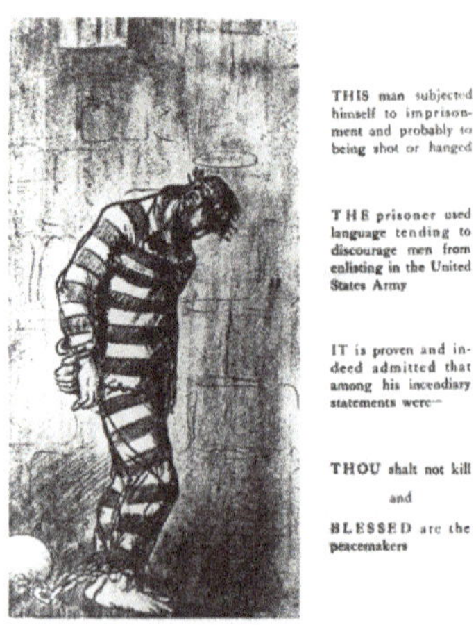

Blessed are the Peacemakers (1917) by George Bellows

Peace Churches

Peace churches are Christian denominations explicitly advocating pacifism. The term "historic peace churches" refers specifically to three church traditions: the Church of the Brethren, the Mennonites (and some other Anabaptists, such as Amish and Hutterites), and the Quakers (Religious Society of Friends). The historic peace churches have, from their origins as far back as the 16th century, always taken the position that Jesus was himself a pacifist who explicitly taught and practiced pacifism, and that his followers must do likewise. Pacifist churches vary on whether physical force can ever be justified in self-defense or protecting others, as many adhere strictly to nonresistance when confronted by violence. But all agree that violence on behalf of a country or a government is prohibited for Christians.

Pentecostal Churches

Jay Beaman's thesis states that 13 of 21, or 62% of American Pentecostal groups formed by 1917 show evidence of being pacifist sometime in their history. Furthermore, Jay Beaman has shown in his thesis that there has been a shift away from pacifism in the American Pentecostal churches to more a style of military support and chaplaincy. The major organisation for Pentecostal Christians who believe in pacifism is the PCPF, the Pentecostal Charismatic Peace Fellowship.

The United Pentecostal Church, the largest Apostolic/Oneness denomination, takes an official stand of conscientious objection: its Articles of Faith read, "We are constrained to declare against participating in combatant service in war, armed insurrection ... aiding or abetting in or the actual destruction of human life. We believe that we can be consistent in serving our Government in certain noncombatant capacities, but not in the bearing of arms."

Other Christian denominations

The shadow of the cross symbolizes the connection between religion and war in *Constantine's Sword (film)*

The Peace Pledge Union was a pacifist organisation from which the Anglican Pacifist Fellowship (APF) later emerged within the Anglican Church. The APF succeeded in gaining ratification of the pacifist position at two successive Lambeth Conferences, but many Anglicans would not regard themselves as pacifists. South African Bishop Desmond Tutu is the most prominent Anglican pacifist. Rowan Williams led an almost united Anglican Church in Britain in opposition to the 2003 Iraq War. In Australia Peter Carnley similarly led a front of bishops opposed to the Government of Australia's involvement in the invasion of Iraq.

The Catholic Worker Movement is concerned with both social justice and pacifist issues, and voiced consistent opposition to the Spanish Civil War and World War II. Many of its early members were imprisoned for their opposition to conscription. Within the Roman Catholic Church, the Pax Christi organisation is the premiere pacifist lobby group. It holds positions similar to APF, and the two organisations are known to work together on ecumenical projects. Within Roman Catholicism there has been a discernible move towards a more pacifist position through the twentieth and early twenty-first centuries. Popes Benedict XV, John XXIII and John Paul II were all vocal in their opposition to specific wars. By taking the name Benedict XVI, some suspected that Joseph Ratzinger would continue the strong emphasis upon nonviolent conflict resolution of his predecessor. However, the Roman Catholic Church officially maintains the legitimacy of Just War, which is rejected by some pacifists.

In the twentieth century there was a notable trend among prominent Roman Catholics towards pacifism. Individuals such as Dorothy Day and Henri Nouwen stand out among them. The monk and mystic Thomas Merton was noted for his commitment to pacifism during the Vietnam War era. Murdered Salvadoran Bishop Oscar Romero was notable for using non-violent resistance tactics and wrote meditative sermons focusing on the power of prayer and peace. School of the Americas Watch was founded by Maryknoll Fr. Roy Bourgeois in 1990 and uses strictly pacifist principles to protest the training of Latin American military officers by United States Army officers at the School of the Americas in the state of Georgia.

The Greek Orthodox Church also tends towards pacifism, though it has accepted defensive warfare through most of its history. However, more recently it took a strong stance towards the war in Lebanon and its large community there refused to take up arms during its civil wars. It also supports dialogue with Islam. In 1998 the Third Pre-conciliar Pan-Orthodox Conference drew up a text on "the contribution of the Orthodox Church to the achievement of peace" emphasizing respect for the human person and the inseparability of peace from justice. The text states in part: "Orthodoxy condemns war in general, for she regards it as a consequence of the evil and sin in the world."

The Southern Baptist Convention has stated in the Baptist Faith and Message, "It is the duty of Christians to seek peace with all men on principles of righteousness. In accordance with the spirit and teachings of Christ they should do all in their power to put an end to war."

The United Methodist Church explicitly supports conscientious objection by its members "as an ethically valid position" while simultaneously allowing for differences of opinion and belief for those who do not object to military service.

Members of the Rastafari Movement's Mansion Nyabinghi are specifically noted for having a large population of Pacifist members, though not all of them are.

Hinduism

Non violence, or ahimsa, is a central part of Hinduism and is one of the fundamental Yamas - self restraints needed to live a proper life. The concept of ahimsa grew gradually within Hinduism, one of the signs being the discouragement of ritual animal sacrifice. Most Hindus today have a vegetarian diet. There are debates on how far the principle of ahimsa applies and if there is such a thing as a "just war".

Jainism

Non-violence, Compassion for all life, human and non-human, is central to Jainism. Human life is valued as a unique, rare opportunity to reach enlightenment. Killing any person, no matter what crime he may have committed, is considered unimaginably terrible. It is a religion that requires monks, from all its sects and traditions, to be vegetarian. Some Indian regions, such as Gujarat, Madhya Pradesh have been strongly influenced by Jains and often the majority of the local Hindus of every denomination are also vegetarian.

Judaism

The attitude of Jews toward pacifism, as with most other aspects of religion, is heavily influenced by the Holocaust which was a program of Nazi Germany to murder every man, woman and child who was Jewish as well as people of other religions who had a Jewish grandparent. As a result, some six million people were exterminated by various means because the Nazis considered them Jews and therefore unworthy of life. In hindsight, there were opportunities for a number of years to defeat Nazi Germany before it could build a military force strong enough to capture and kill most of the Jews of Europe. Forces which deterred the democracies from acting to stop Hitler early on and at a much, much lower cost in human life were pacifism, appeasement, and isolationism. People who offered no resistance to the Nazis enabled them to carry out their oppression and aggression, costing tens of millions of lives in World War II. Had pacifism gained more support, the Allies might have lost the war and virtually all Jews would have been killed.

The Jewish Peace Fellowship is a New-York based nonprofit, nondenominational organization set up to provide a Jewish voice in the peace movement. The organization was founded in 1941 in order to support Jewish conscientious objectors who sought exemption from combatant military service. It is affiliated to the International Fellowship of Reconciliation. The small Neturei Karta group of anti-Zionist, ultra-orthodox Jews, supposedly take a pacifist line, saying that "Jews are not allowed to dominate, kill, harm or demean another people and are not allowed to have anything to do with the Zionist enterprise, their political meddling and their wars.". However, the Neturei Karta group do support groups such as Hezbollah and Hamas that are violent towards Israel. The Hebrew Bible is full of examples when Jews were told to go and war against enemy lands or within the Israelite community as well as instances where God, as destroyer and protector, goes to war for non-participant Jews. The Holocaust Remembrance Day (called Yom Hashoah in Hebrew) is a day a remembrance for many Jews as they honor those who fought to end the Hitler government which starved, shot, gassed and burned over six million Jews to death. It is observed on the day corresponding to the 27th day of the month of Nisan on the Hebrew calendar.

Raëlism

Non-violence is an important doctrine within Raëlism. The founder of this religion Rael has said "*The one holding the weapon is as responsible as the one giving the orders*". Other Rael statements include "*even if the Elohim asked them to kill someone they should refuse*".

Government and Political Movements

While many governments have tolerated pacifist views and even accommodated pacifists' refusal to fight in wars, others at times have outlawed pacifist and anti-war activity. In 1918, The United States Congress passed the Sedition Act of 1918. During the periods between World Wars I and World War II, pacifist literature and public advocacy was banned in Italy under Benito Mussolini, Germany after the rise of Adolf Hitler, Spain under Francisco Franco, and the Soviet Union under Joseph Stalin. In these nations, pacifism was denounced as cowardice; indeed, Mussolini referred to pacifist writings as the "propaganda of cowardice".

Remarque's anti-war novel *All Quiet on the Western Front* was banned and burned by war-glorifying Nazis

Today, the United States requires that all young men register for selective service but does not allow them to be classified as conscientious objectors unless they are drafted in some future re-instatement of the draft, allowing them to be discharged or transferred to noncombatant status. Some European governments like Switzerland, Greece, Norway and Germany offer civilian service. However, even during periods of peace, many pacifists still refuse to register for or report for military duty, risking criminal charges.

Anti-war and "pacifist" political parties seeking to win elections may moderate their demands, calling for de-escalation or major arms reduction rather than the outright disarmament which is advocated by many pacifists. Green parties list "non-violence" and "decentralization" towards anarchist co-operatives or minimalist village government as two of their ten key values. However, in power, Greens often compromise. The German Greens in the cabinet of Social Democrat Gerhard Schröder supported an intervention by German troops in Afghanistan in 2001 if that they hosted the peace conference in Berlin. However, during the 2002 election Greens forced Schröder to swear that no German troops would invade Iraq.

The controversial democratic peace theory holds that liberal democracies have never (or rarely) made war on one another and that lesser conflicts and internal violence are rare between and within democracies. It also argues that the growth in the number of democratic states will, in the not so distant future, end warfare.

Some pacifists and multilateralists are in favor of international criminal law as means to prevent and control international aggression. The International Criminal Court has jurisdiction over war crimes, but the crime of aggression has yet to be clearly defined in international law.

The Italian Constitution enforces a mild pacifist character on the Italian Republic, as Article 11 states that "Italy repudiates war as an instrument offending the liberty of the peoples and as a means for settling international disputes" Similarly, Articles 24, 25 and 26 of the German Constitution (1949), Alinea 15 of the French Constitution (1946), Article 20 of the Danish Constitution (1953), Article 9 of the Japanese Constitution (1947) and several other mostly European constitutions correspond to the United Nations Charter by rejecting the institution of war in favour of collective security and peaceful cooperation.

Pacifism and Abstention from Political Activity

However, some pacifists, such as the Christian anarchist Leo Tolstoy and autarchist Robert LeFevre, consider the state a form of warfare. In addition, for doctrinal reason that a manmade government is inferior to divine governance and law, many pacifist-identified religions/religious sects also refrain from political activity altogether, including the Anabaptists, Jehovah's Witnesses and Mandaeans. This means that such groups refuse to participate in government office or serve under an oath to a government.

Anarcho-pacifism

Henry David Thoreau, early proponent of anarcho-pacifism

Anarcho-pacifism (also pacifist anarchism or anarchist pacifism) is a form of anarchism which completely rejects the use of violence in any form for any purpose. The main precedent was Henry David Thoreau who through his work Civil Disobedience influenced the advocacy of both Leo Tolstoy and Mohandas Gandhi for nonviolent resistance. As a global movement, Anarchist pacifism emerged shortly before World War II in the Netherlands, Great Britain and the United States and was a strong presence in the subsequent campaigns for nuclear disarmament.

Violence has always been controversial in anarchism. While many anarchists during the 19th century embraced propaganda of the deed, Leo Tolstoy and other anarcho-pacifists directly opposed violence as a means for change. He argued that anarchism must by nature be nonviolent since it is, by definition, opposition to coercion and force and since the state is inherently violent, meaningful pacifism must likewise be anarchistic. His philosophy was cited as a major inspiration by Mohandas Gandhi, an Indian independence leader and pacifist who self-identified as an anarchist. Ferdinand Domela Nieuwenhuis was also instrumental in establishing the pacifist trend within the anarchist movement. In France anti-militarism appeared strongly in individualist anarchist circles as Émile Armand founded "Ligue Antimilitariste" in 1902 with Albert Libertad and George Mathias Paraf-Javal.

Opposition to Military Taxation

Many pacifists who would be conscientious objectors to military service are also opposed to paying taxes to fund the military. In the United States, The National Campaign for a Peace Tax Fund works to pass a national law to allow conscientious objectors to redirect their tax money to be used only for non-military purposes.

Criticism

One common argument against pacifism is the possibility of using violence to prevent further acts of violence (and reduce the "net-sum" of violence). This argument hinges on consequentialism: an otherwise morally objectionable action can be justified if it results in a positive outcome. For example, either violent rebellion, or foreign nations sending in troops to end a dictator's violent oppression may save millions of lives, even if many thousands died in the war. Those pacifists who base their beliefs on deontological grounds would oppose such violent action. Others would oppose organized military responses but support individual and small group self-defense against specific attacks if initiated by the dictator's forces. Pacifists may argue that military action could be justified should it subsequently advance the general cause of peace.

Still more pacifists would argue that a nonviolent reaction may not save lives immediately but would in the long run. The acceptance of violence for any reason makes it easier to use in other situations. Learning and committing to pacifism helps to send a message that violence is, in fact, not the most effective way. It can also help people to think more creatively and find more effective ways to stop violence without more violence.

In light of the common criticism of pacifism as not offering a clear alternative policy, one approach to finding "more effective ways" has been the attempt to develop the idea of "defence by civil resistance", also called "social defence". This idea, which is not necessarily dependent on acceptance of pacifist beliefs, is based on relying on nonviolent resistance against possible threats, whether external (such as invasion) or internal (such as coup d'état).

Jewish armed resistance against the Nazis during World War II

There have been some works on this topic, including by Adam Roberts and Gene Sharp. However, no country has adopted this approach as the sole basis of its defence.

Japanese, Italian and Nazi aggression that precipitated World War II often is cited as an argument against pacifism. If these forces had not been challenged and defeated militarily, the argument goes, many more people would have died under their oppressive rule. Adolf Hitler told the British Foreign Secretary Lord Halifax in 1937 that the British should "shoot Gandhi, and if this doesn't suffice to reduce them to submission, shoot a dozen leading members of the Congress, and if that doesn't suffice shoot 200, and so on, as you make it clear that you mean business."

Adolf Hitler noted in his Second Book: "Later, the attempt to adapt the living space to increased population turned into unmotivated wars of conquest, which in their very lack of motivation contained the germ of the subsequent reaction. Pacifism is the answer to it. Pacifism has existed in the world ever since there have been wars whose meaning no longer lay in the conquest of territory for a Folk's sustenance. Since then it has been war's eternal companion. It will again disappear as soon as war ceases to be an instrument of booty hungry or power hungry individuals or nations, and as soon as it again becomes the ultimate weapon with which a Folk fights for its daily bread."

Hermann Göring described, during an interview at the Nuremberg Trials, how denouncing and outlawing pacifism was an important part of the Nazis' seizure of power: "The people can always be brought to the bidding of the leaders. That is easy. All you have to do is tell them they are being attacked and denounce the pacifists for lack of patriotism and exposing the country to danger. It works the same way in any country."

Some commentators on the most nonviolent forms of pacifism, including Jan Narveson, argue that such pacifism is a self-contradictory doctrine. Narveson claims that everyone has rights and corresponding responsibilities not to violate others' rights. Since pacifists give up their ability to protect themselves from violation of their right not to be harmed, then other people thus have no corresponding responsibility, thus creating a paradox of rights. Narveson said that "the prevention of infractions of that right is precisely what one has a right to when one has a right at all". Narveson then discusses how rational persuasion is a good but often inadequate method of discouraging an aggressor. He considers that everyone has the right to use any means necessary to prevent deprivation of their civil liberties and force could be necessary.

Peace Movement

A peace movement is a social movement that seeks to achieve ideals such as the ending of a particular war (or all wars), minimize inter-human violence in a particular place or type of situation, and is often linked to the goal of achieving world peace. Means to achieve these ends include advocacy of pacifism, non-violent resistance, diplomacy, boycotts, peace camps, moral purchasing, supporting anti-war political candidates, legislation to remove the profit from government contracts to the Military–industrial complex, banning guns, creating open government and transparency tools, direct democracy, supporting Whistleblowers who expose War-Crimes or conspiracies to create wars, demonstrations, and national political lobbying groups to create legislation. The political cooperative is an example of an organization that seeks to merge all peace movement organizations and green organizations, which may have some diverse goals, but all of whom have the common goal of peace and humane sustainability. A concern of some peace activists is the challenge of attaining peace when those that oppose it often use violence as their means of communication and empowerment.

An Australian anti-conscription poster from World War One

Cover of *Die Friedens-Warte*, a German journal of the peace movement, issue #11, 1913

Some people refer to the global loose affiliation of activists and political interests as having a shared purpose and this constituting a single movement, "*the* peace movement", an all encompassing "anti-war movement". Seen this way, the two are often indistinguishable and constitute a loose, responsive, event-driven collaboration between groups with motivations as diverse as humanism, environmentalism, veganism, anti-racism, anti-sexism, decentralization, hospitality, ideology, theology, and faith.

The Ideal of Peace

There are different ideas over what "peace" is (or should be), which results in a plurality of movements seeking diverse ideals of peace. Particularly, "anti-war" movements often have short-term goals, while peace movements advocate an ongoing life-style and proactive government policy.

It is often not clear whether a movement or a particular protest is against war in general, as in pacifism, or against one's own government's participation in a war. Indeed, some observers feel that this lack of clarity or long term continuity has represented a key part of the strategy of those seeking to end a war, e.g., the Vietnam War.

Global protests against the U.S. invasion of Iraq in early 2003 are an example of a more specific, short term and loosely affiliated single-issue "movement" —with relatively scattered ideological priorities, ranging from absolutist pacifism to Islamism and Anti-Americanism. Nonetheless, some of those who are involved in several such short term movements and build up trust relationships with others within them, do tend to eventually join more global or long-term movements.

1969 Hungarian stamp commemorating the 20th anniversary of peace movement, denomination: 1 Forint

By contrast, some elements of the global peace movement seek to guarantee health security by ending war and assuring what they see as basic human rights including the right of all people to have access to air, water, food, shelter and health care. A number of activists seek social justice in the form of equal protection under the law and equal opportunity under the law for groups that have previously been disenfranchised.

The Peace movement is primarily characterized by a belief that humans should not wage war on each other or engage in violent ethnic cleansings over language, race or natural resources or ethical conflict over religion or ideology. Long-term opponents of war preparations are primarily characterized by a belief that military power is not the equivalent of justice.

The Peace movement tends to oppose the proliferation of dangerous technologies and weapons of mass destruction, in particular nuclear weapons and biological warfare. Moreover, many object to the export of weapons including hand-held machine guns and grenades by leading economic nations to lesser developed nations. Some, like SIPRI, have voiced special concern that artificial intelligence, molecular engineering, genetics and proteomics have even more vast destructive potential. Thus there is intersection between peace movement elements and Neo-Luddites or primitivism, but also with the more mainstream technology critics such as the Green parties, Greenpeace and the ecology movement they are part of.

It is one of several movements that led to the formation of Green party political associations in many democratic countries near the end of the 20th century. The peace movement has a very strong influence in some countries' green parties, such as in Germany, perhaps reflecting that country's negative experiences with militarism in the 20th century.

History

Quakers and the Enlightenment

Beginning in the 16th century, the Protestant Reformation gave rise to a variety of new Christian sects, including the historic peace churches. Foremost among them were the Religious Society of Friends (Quakers), Amish, Mennonites and Church of the Brethren. The Quakers were prominent advocates of pacifism, who as early as 1660 had repudiated violence in all forms and adhered to a strictly pacifist interpretation of Christianity. Throughout the many 18th century wars in which Britain participated, the Quakers maintained a principled commitment not to serve in the army and militia or even to pay the alternative £10 fine.

The first Peace movements were products of two strands of thought that coalesced at the end of the 18th century. One, rooted in the secular Enlightenment, promoted peace as the rational antidote to the world's ills, while the other was a part of the evangelical religious revival that had played an important part in the campaign for the abolition of slavery. Representatives of the former, included Jean-Jacques Rousseau, in *Extrait du Projet de Paix Perpetuelle de Monsieur l'Abbe Saint-Pierre* (1756), Immanuel Kant, in his *Thoughts on Perpetual Peace.* and Jeremy Bentham who proposed the formation of a peace association in 1789. Representative of the latter, was William Wilberforce who thought that strict limits should be imposed on British involvement in the French Revolutionary War based on Christian ideals of peace and brotherhood.

Peace Movements

During the period of the Napoleonic Wars, although no formal peace movement was established until the end of hostilities, a significant peace movement animated by universalist ideals did emerge, due to the perception of Britain fighting in a reactionary role and the increasingly visible impact of the war on the welfare of the nation in the form of higher taxation levels and high casualty rates. Sixteen peace petitions to Parliament were signed by members of the public, anti-war and anti-Pitt demonstrations convened and peace literature was widely published and disseminated.

The first peace movements appeared in 1815–16. In the United States the first such movement was the New York Peace Society, founded in 1815 by the theologian David Low Dodge, and the Massachusetts Peace Society. It became an active organization, holding regular weekly meetings, and producing literature which was spread as far as Gibraltar and Malta, describing the horrors of war and advocating pacificism on Christian grounds. The London Peace Society (also known as the Society for the Promotion of Permanent and Universal Peace) was formed in 1816 to promote permanent and universal peace by the philanthropist William Allen. In the 1840s, British women formed "Olive Leaf Circles", groups of around 15 to 20 women, to discuss and promote pacifist ideas.

The peace movement began to grow in influence by the mid nineteenth century. The London Peace Society, under the initiative of American consul to Birmingham Elihu Burritt and the reverend Henry Richard, convened the first International Peace Congress in London in 1843. The congress decided on two aims: the ideal of peaceable arbitration in the affairs of nations and the creation of an international institution to achieve that. Richard became the secretary of the Peace Society in 1850 on a full-time basis, a position which he would keep for the next 40 years, earning himself a reputation as the 'Apostle of Peace'. He helped secure one of the earliest victories for the peace movement by securing a commitment from the Great Powers in the Treaty of Paris (1856) at the end of the Crimean War, in favour of arbitration. On the European continent, wracked by social upheaval, the first peace congress was held in Brussels in 1848 followed by Paris a year later.

After experiencing a recession in support due to the resurgence of militarism during the American Civil War and Crimean War, the movement began to spread across Europe and began to infiltrate the new working class socialist movements. In 1870, Randal Cremer formed the Workman's Peace Association in London. Cremer, alongside the French economist Frédéric Passy was also the founding father of the first international organisation for the arbitration of conflicts in 1889, the Inter-Parliamentary Union. The National Peace Council was founded in after the 17th Universal Peace Congress in London (July August 1908).

Nonviolent Resistance

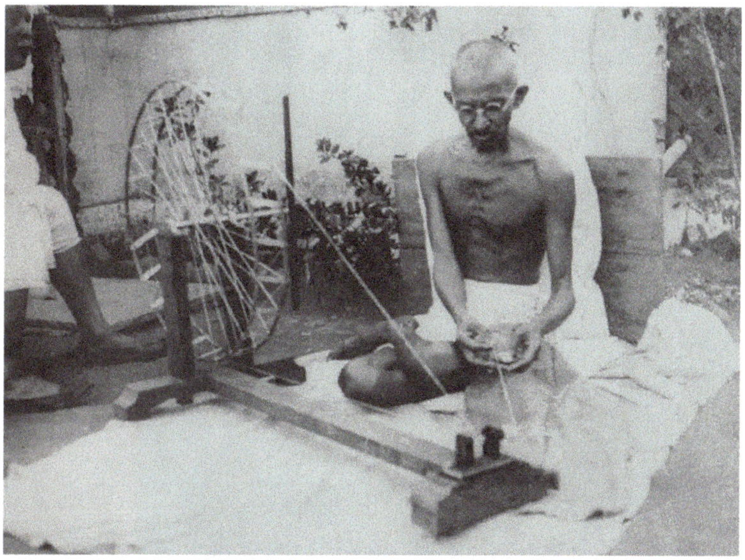

Mahatma Gandhi, leader of the Indian independence movement and advocate of nonviolent resistance.

Mahatma Gandhi (1869–1948) of India was one of the most influential spokesman for peace and non-violence in the 20th century. *Gandhism* comprises the ideas and principles Gandhi promoted. Of central importance is nonviolent resistance. M.M. Sankhdher argues that Gandhism is not a systematic position in metaphysics or in political philosophy. Rather, it is a political creed, an economic doctrine, a religious outlook, a moral precept, and especially, a humanitarian world view. It is an effort not to systematize wisdom but to transform society and is based on an undying faith in the goodness of human nature. Gandhi was strongly influenced by the pacifist ideas of Russian novelist Leo Tolstoy. In 1908 Tolstoy wrote *A Letter to a Hindu,* which said that only by using love as a weapon through passive resistance could the Indian people overthrow colonial rule. In 1909, Gandhi and Tolstoy began a correspondence regarding practical and theological applications of non-violence. Gandhi saw himself a disciple of Tolstoy, for they agreed regarding opposition to state authority and colonialism; both hated violence and preached non-resistance. However, they differed sharply on political strategy. Gandhi called for political involvement; he was a nationalist and was prepared to use nonviolent force. He was also willing to compromise.

Gandhi was the first to apply the principle of nonviolence on a large scale. The concept of nonviolence (*ahimsa*) and nonresistance has a long history in Indian religious thought and has had many revivals in Hindu, Buddhist, Jain, Jewish and Christian contexts. Gandhi explains his philosophy and way of life in his autobiography *The Story of My Experiments with Truth*. Some of his other remarks were widely quoted, such as "There are many causes that I am prepared to die for but no causes that I am prepared to kill for."

Gandhi realized later that this level of nonviolence required incredible faith and courage, which he believed everyone did not possess. He therefore advised that everyone need not keep to nonviolence, especially if it were used as a cover for cowardice, saying, "where there is only a choice between cowardice and violence, I would advise violence."

Gandhi came under political fire for his criticism of those who attempted to achieve independence through more violent means. Gandhi responded, "There was a time when people listened to me because I showed them how to give fight to the British without arms when they had no arms but today I am told that my non-violence can be of no avail against the [Hindu–Moslem riots] and, therefore, people should arm themselves for self-defense."

Gandhi's views came under heavy criticism in Britain when it was under attack from Nazi Germany. He told the British people in 1940, "I would like you to lay down the arms you have as being useless for saving you or humanity. You will invite Herr Hitler and Signor Mussolini to take what they want of the countries you call your possessions... If these gentlemen choose to occupy your homes, you will vacate them. If they do not give you free passage out, you will allow yourselves, man, woman, and child, to be slaughtered, but you will refuse to owe allegiance to them."

Challenge of World War I

The Deserter (1916) by Boardman Robinson.

Although the onset of the First World War was generally greeted with enthusiastic patriotism across Europe, peace groups were still active in condemning the war. Many socialist groups and movements were antimilitarist, arguing that war by its nature was a type of governmental coercion of the working class for the benefit of capitalist elites.

In 1915 the League of Nations Society was formed by British liberal leaders to promote a strong international organisation that could enforce the peaceful resolution of conflict. Later that year the League to Enforce Peace was established in America to promote similar goals. Hamilton Holt published an editorial in his New York City weekly magazine the *Independent* called "The Way to Disarm: A Practical Proposal" on September 28, 1914. It called for an international organization to agree upon the arbitration of disputes and to guarantee the territorial integrity of its members by maintaining military forces sufficient to defeat those of any non-member. The ensuing debate among prominent internationalists modified Holt's plan to align it more closely with proposals offered in Great Britain by Viscount James Bryce, a former ambassador from the UK to the US. These and other initiatives were pivotal in the change in attitudes that gave birth to the League of Nations after the war.

Some of the many groups that protested against the war, as well as the traditional peace churches, were the Woman's Peace Party (which was organized in 1915 and led by noted reformer Jane Addams), the International Committee of Women for Permanent Peace (ICWPP) (also organized in 1915), the American Union Against Militarism, the Fellowship of Reconciliation, and the American Friends Service Committee. Jeannette Rankin, the first woman elected to Congress, was another fierce advocate of pacifism, the only person to vote no to America's entrance into both World Wars.

Interwar Period

The immense loss of life during the war, for what became regarded as futile reasons, caused a sea-change in public attitudes to militarism. Organisations formed in this period included the War Resisters' International the Women's International League for Peace and Freedom, the No More War Movement and the Peace Pledge Union (PPU). The League of Nations also convened several disarmament conferences in the inter-war period such as the Geneva Conference.

Pacifism and revulsion with war were very popular sentiments in 1920s Britain. A stream of novels and poems on the theme of the futility of war and the slaughter of the youth by old fools were published, including, Death of a Hero by Richard Aldington, Erich Remarque's translated All Quiet on the Western Front and Beverley Nichols's expose, *Cry Havoc*. A debate at the University of Oxford in 1933 on the motion 'one must fight for King and country' captured the changed mood when the motion was resoundingly defeated. Dick Sheppard established the Peace Pledge Union in 1934 totally renouncing war and aggression. The idea of collective security was also popular; instead of outright pacifism the public generally exhibited a determination to stand up to aggression, but preferably with the use of economic sanctions and multilateral negotiations.

The Spanish Civil War proved a major test for international pacifism, and the work of pacifist organisations (such as War Resisters' International and the Fellowship of Reconciliation) and individuals (such as José Brocca and Amparo Poch). Shortly after the war ended, Simone Weil, despite having volunteered for service on the republican side, went on to publish *The Iliad or the Poem of Force*, a work that has been described as a pacifist manifesto. In response to the threat of fascism, some pacifist thinkers, such as Richard B. Gregg, devised plans for a campaign of nonviolent resistance in the event of a fascist invasion or takeover.

World War II

With the start of World War II, pacifist and anti-war sentiment declined in nations affected by war. Even the communist-controlled American Peace Mobilization reversed its anti-war activism once Germany invaded the Soviet Union in 1941. After the Japanese attack on Pearl Harbor, mainstream isolationist groups like the America First Committee, declined, but many smaller religious and socialist groups continued their opposition to war. Bertrand Russell argued that the necessity of defeating Adolf Hitler and the Nazis was a unique circumstance where war was not the worst of the possible evils; he called his position *relative pacifism*. H. G. Wells, who had joked after the armistice ending World War I that the British had suffered more from the war than they would have from submission to Germany, urged in 1941 a large-scale British offensive on the continent of Europe to combat Hitler and Nazism. Similarly Albert Einstein wrote: "I loathe all armies and any kind of violence; yet I'm firmly convinced that at present these hateful weapons offer the only effective protection."

The French pacifists André and Magda Trocmé helped conceal hundreds of Jews fleeing the Nazis in the village of Le Chambon-sur-Lignon. After the war, the Trocmés were declared Righteous Among the Nations.

Pacifists in the Third Reich were dealt with harshly; German pacifist Carl von Ossietzky, and Olaf Kullmann, a Norwegian pacifist active during the Nazi occupation, were both imprisoned in concentration camps and died as a result of their mistreatment there. Austrian farmer Franz Jägerstätter was executed in 1943 for refusing to serve in the Wehrmacht.

There were conscientious objectors and war tax resisters in both World War I and World War II. The United States government allowed sincere objectors to serve in noncombatant military roles. However, those draft resisters who refused any cooperation with the war effort often spent much of each war in federal prisons. During World War II, pacifist leaders like Dorothy Day and Ammon Hennacy of the Catholic Worker Movement urged young Americans not to enlist in military service.

Since World War II, the peace movement has become widespread throughout the world, and its previously radical doctrines are now a part of mainstream political discourse.

Anti-nuclear Movement

Peace movements emerged in Japan and in 1954 they converged to form a unified "Japanese Council Against Atomic and Hydrogen Bombs". Japanese opposition to the Pacific nuclear weapons tests was widespread, and "an estimated 35 million signatures were collected on petitions calling for bans on nuclear weapons".

In the United Kingdom, the Campaign for Nuclear Disarmament held an inaugural public meeting at Central Hall, Westminster, on 17 February 1958, attended by five thousand people. After the meeting a few hundred left to demonstrate at Downing Street.

CND's declared policies were the unconditional renunciation of the use, production of or dependence upon nuclear weapons by Britain and the bringing about of a general disarmament convention. In the meantime, Britain should halt the flight of planes armed with nuclear weapons, end nuclear testing, not proceed with missile bases and not provide nuclear weapons to any other country.

A nuclear fireball lights up the night in a United States nuclear weapons test.

The first Aldermaston March was organised by the CND and took place at Easter 1958, when several thousand people marched for four days from Trafalgar Square, London, to the Atomic Weapons Research Establishment close to Aldermaston in Berkshire, England, to demonstrate their opposition to nuclear weapons. The Aldermaston marches continued into the late 1960s when tens of thousands of people took part in the four-day marches. The formation of CND tapped widespread popular fear and opposition to nuclear weapons following the development of the first hydrogen bomb, and in the late 1950s and early 1960s anti-nuclear marches attracted large followings, especially to the annual Aldermaston march at Easter.

Popular opposition to nuclear weapons produced a Labour Party resolution for unilateral nuclear disarmament at the 1960 Party Conference, but it was overturned the following year and did not appear on later agendas. This experience disillusioned many anti-nuclear protesters with the Labour Party, in whom they had previously put their hopes. Subsequently, there was a strong anti-parliamentary current in the British peace movement, and it has been argued that during the 1960s anarchism became as influential as socialism.

Two years after the formation of CND Bertrand Russell, its president, resigned to form the Committee of 100, which was to undertake civil disobedience in the form of sit-down demonstrations in central London and at nuclear bases around the UK. Russell said that these were needed because the press had grown indifferent to CND and because large-scale direct action could force the government to change its policy. A hundred prominent people, many in the arts, put their names to the organisation. Very large numbers of demonstrators were essential to this strategy, but the violence of the police, the arrest and imprisonment of demonstrators, and pre-emptive arrests for conspiracy made support dwindle rapidly. Although several eminent people took part in sit-down demonstrations (including Russell, whose imprisonment at the age of 89 was widely reported) many of the 100 signatories were inactive.

As the Committee of 100 had a non-hierarchical structure and no formal membership, many local groups sprang up calling themselves Committee of 100. This helped the promulgation of civil disobedience but it produced policy confusion and, as the decade progressed, Committee of 100 groups engaged in actions on many social issues not directly related to war and peace.

Women Strike for Peace during the Cuban Missile Crisis

In the United States, in 1961, at the height of the Cold War, about 50,000 women brought together by Women Strike for Peace marched in 60 cities in the United States to demonstrate against nuclear weapons. It was the largest national women's peace protest of the 20th century.

In 1958, Linus Pauling and his wife presented the United Nations with the petition signed by more than 11,000 scientists calling for an end to nuclear-weapon testing. The "Baby Tooth Survey," headed by Dr Louise Reiss, demonstrated conclusively in 1961 that above-ground nuclear testing posed significant public health risks in the form of radioactive fallout spread primarily via milk from cows that had ingested contaminated grass. Public pressure and the research results subsequently led to a moratorium on above-ground nuclear weapons testing, followed by the Partial Test Ban Treaty, signed in 1963 by John F. Kennedy, Nikita Khrushchev and Harold Macmillan. On the day that the treaty went into force, the Nobel Prize Committee awarded Pauling the Nobel Peace Prize, describing him as "Linus Carl Pauling, who ever since 1946 has campaigned ceaselessly, not only against nuclear weapons tests, not only against the spread of these armaments, not only against their very use, but against all warfare as a means of solving international conflicts." Pauling started the International League of Humanists in 1974. He was president of the scientific advisory board of the World Union for Protection of Life and also one of the signatories of the Dubrovnik-Philadelphia Statement.

On June 12, 1982, one million people demonstrated in New York City's Central Park against nuclear weapons and for an end to the cold war arms race. It was the largest anti-nuclear protest and the largest political demonstration in American history. International Day of Nuclear Disarmament protests were held on June 20, 1983 at 50 sites across the United States. In 1986, hundreds of people walked from Los Angeles to Washington DC in the Great Peace

March for Global Nuclear Disarmament. There were many Nevada Desert Experience protests and peace camps at the Nevada Test Site during the 1980s and 1990s.

On May 1, 2005, 40,000 anti-nuclear/anti-war protesters marched past the United Nations in New York, 60 years after the atomic bombings of Hiroshima and Nagasaki. This was the largest anti-nuclear rally in the U.S. for several decades. In Britain, there were many protests about the government's proposal to replace the aging Trident weapons system with a newer model. The largest protest had 100,000 participants and, according to polls, 59 percent of the public opposed the move.

The International Conference on Nuclear Disarmament took place in Oslo in February 2008, and was organized by The Government of Norway, the Nuclear Threat Initiative and the Hoover Institute. The Conference was entitled *Achieving the Vision of a World Free of Nuclear Weapons* and had the purpose of building consensus between nuclear weapon states and non-nuclear weapon states in relation to the Nuclear Non-proliferation Treaty.

In May 2010, some 25,000 people, including members of peace organizations and 1945 atomic bomb survivors, marched for about two kilometers from downtown New York to the United Nations headquarters, calling for the elimination of nuclear weapons.

Vietnam War Protests

Protests against the Vietnam War in Washington, D.C., on October 21, 1967

The peace movement began in the 1960s in the United States in opposition to U.S. involvement in the Vietnam War. Some advocates within this movement advocated a unilateral withdrawal of U.S. forces from South Vietnam.

The first U.S. anti-Vietnam protest was led in 1962 by Sam Marcy, founder of Workers World Party, a demonstration whose importance was noted by Ho Chi Minh in an interview published in the National Guardian newspaper.

Opposition to the Vietnam War tended to unite groups opposed to U.S. anti-communism, imperialism and colonialism and, for those involved with the New Left, capitalism itself, such as the Catholic Worker Movement. Others, such as Stephen Spiro opposed the war based on the theory of Just War.

In 1965 the movement began to gain national prominence. Provocative actions by police and by protesters turned anti-war demonstrations in Chicago at the 1968 Democratic National Convention into a riot. Explosive news reports of American military abuses, such as the 1968 My Lai Massacre, brought new attention and support to the anti-war movement bringing it to its height. The movement continued to prosper over the span of the conflict.

High-profile opposition to the Vietnam war turned to street protests in an effort to turn U.S. political opinion against the war. The protests gained momentum from the Civil Rights Movement that had organized to oppose segregation laws, which had laid a foundation of theory and infrastructure on which the anti-war movement grew. Protests were fueled by a growing network of independently published newspapers (known as "underground papers") and the timely advent of large venue rock'n'roll festivals such as Woodstock and Grateful Dead shows, attracting younger people in search of generational togetherness. The movement progressed from college campuses to middle-class suburbs, government institutions, and labor unions.

Peace Movements by Country

Sweden: Stockholm's *Peace and Love Rally* through the south side of the city drew hundreds of marchers and celebrants in May 2015

Canada

Agnes Macphail, a Canadian pacifist, was the first woman to be elected to the Canadian House of Commons. Macphail objected to the Royal Military College of Canada in 1931 on pacific grounds. Macphail was also the first Canadian woman delegate to the League of Nations, where she worked with the World Disarmament Committee. Although a pacifist, she voted for Canada to enter World War II.

The Canadian Peace Congress (1949–1990) was a leading organizer in the peace movement for many years, particularly when it was under the leadership of James Gareth Endicott who was its president until 1971.

Currently, Canada has a diverse peace movement, with coalitions and networks in many cities, towns and regions. The largest cross-country umbrella coalition is the Canadian Peace Alliance, whose 140 member groups include large city-based coalitions, small grassroots groups, national and local unions, faith, environmental, and student groups, with a combined membership of over 4 million Canadians. The Canadian Peace Alliance has been a leading voice, along with its member groups opposing the "War on Terror". In particular, the CPA opposes Canada's participation in the war in Afghanistan and Canadian complicity in what it views as misguided and destructive U.S. foreign policy.

Canada has also been home to a growing movement of Palestinian solidarity, marked by an increasing number of grassroots Jewish groups opposed to Israel's policies, in many cases likening them to Apartheid, war crimes, and ethnic cleansing.

Germany

During the Cold War (1947–89), the West German peace movement concentrated on the abolition of nuclear technology, particularly weapons, from West Germany and Europe. Most activists stridently attacked both the United States and Soviet Union. Conservative critics repeatedly warned it was infiltrated by agents from the East German secret police, the Stasi.

After 1989, the cause of peace was espoused by Green parties across Europe. It sometimes exercised significant influence over policy, e.g., as during 2002 when the German Greens influenced German Chancellor Gerhard Schröder to oppose involvement in Iraq. The Greens controlled of the German Foreign Ministry under Joschka Fischer (a Green and the single most popular politician in Germany at the time). Fischer sought to limit German involvement in the War on Terrorism; he joined with French President Jacques Chirac whose opposition in the UN Security Council was decisive in limiting support for the U.S. plan to invade Iraq.

Israel

The Israeli–Palestinian conflict and Arab–Israeli conflict have existed since the inception of Zionism, and especially since the 1948 formation of the state of Israel, and the 1967 Six-Day War. The mainstream peace movement in Israel is Peace Now (*Shalom Akhshav*), whose supporters tend to vote for the Labour Party or Meretz.

Peace Now was founded in the aftermath of Egyptian President Anwar Sadat's historic visit to Jerusalem, when many people felt that the chance for peace might be missed. PM Begin acknowledged that the Peace Now rally in Tel Aviv at the eve of his departure for the Camp David Summit with Presidents Sadat and Carter—drawing a crowd of 100,000, the largest peace rally in Israel until then—had a part in his decision to withdraw from Sinai and dismantle Israeli settlements there. Peace Now supported Begin for a time, and hailed him as a peace-maker, but turned against him when withdrawal from Sinai was accompanied by an accelerated campaign of land confiscation and settlement building in the West Bank.

Peace Now advocates a negotiated peace with the Palestinians. Originally this was worded vaguely, with no definition of who "the Palestinians" are and who represents them. Peace Now was quite tardy in joining the dialogue with the PLO, started by such groups as the Israeli Council for Israeli-Palestinian Peace and the Hadash communist party. Only in 1988 did Peace Now accept that the PLO is the body regarded by the Palestinians themselves as their representative.

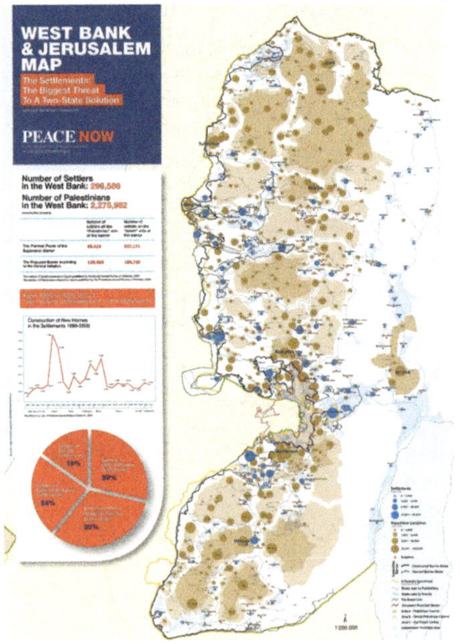

A map showing settlements in the West Bank, produced by Peace Now.

During the first Intifada, Peace Now held numerous protests and rallies to protest the Israeli army and call for a negotiated withdrawal from the Palestinian territories. At the time Peace Now strongly targeted then for Defence Minister Yitzhak Rabin for his order to "break the bones of Palestinian stone-throwers". However, after Rabin became Prime Minister, signed the Oslo Agreement and shook Yasser Arafat's hand on the White House lawn, Peace Now strongly supported him and mobilized public support for him. Peace Now had a central role in the November 4, 1995 rally after which Rabin was assassinated.

Since then the annual Rabin memorial rallies, held every year at the beginning of November, have become the main event of the Israeli Peace Movement. While officially organized by the Rabin Family Foundation, Peace Now presence in these annual rallies is always conspicuous. Nowadays, Peace Now is especially known for its struggle against the expansion of settlement outposts on the West Bank.

Gush Shalom, the Israeli Peace Bloc, is a radical movement to the left of Peace Now. In its present name and structure, Gush Shalom grew out of the *Jewish-Arab Committee Against Deportations*, which protested the deportation without trial of 415 Palestinian Islamic activists to Lebanon in December 1992, and erected a protest tent in front of the prime minister's office in Jerusalem for two months—until the government consented to let the deportees return. Members then decided to continue as a general peace movement with a program strongly opposing the occupation and advocating the creation of an independent Palestine side-by-side with Israel in its pre-1967 borders ("The Green Line") and with an undivided Jerusalem serving as the capital of both states.

While existing under the name Gush Shalom only since 1992, this movement is in fact the lineal descendant of various groups, movements and action committees that espoused much the same program since 1967, and that occupied the same space on the political scene. In particular, Gush Shalom is the descendant of the *Israeli Council for Israeli-Palestinian Peace* (ICIPP), which was founded in 1975. The ICIPP founders included: a group of dissidents from the Israeli establishment, among them were Major-General Mattityahu Peled, who was member of the IDF General Staff during the 1967 Six Day War and after being dishcarged from the army in 1969 turned increasingly in the direction of the left; Dr. Ya'akov Arnon, a well-known economist who headed the Zionist Federation in the Netherlands before coming to Israel in 1948, and was for many years director-general of the Israeli Ministry of Finance and afterwards chaired the Board of Directors of the Israeli Electricity Company; and Aryeh Eliav, who was secretary-general of the Labour Party until he broke with the then PM Golda Meir over the issue of whether or not a Palestinian People existed and had national rights.

These three and some two hundred more people became radicalised and came to the conclusion that "arrogance was a threat to Israel's future and that dialogue with the Palestinians must be opened." They came together with a group of younger, grassroots peace activists who had been active against Israeli occupation since 1967. The bridge between the two groups was Uri Avnery, a well known muckraking journalist who had been member of the Knesset (Israeli Parliament) between 1965 and 1973, at the head of his own radical one-man party.

The main achievement of the ICIPP was the opening of dialogue with the Palestine Liberation Organization (PLO), with the aim of convincing the Israelis of the need to talk and reach a peace deal with the Palestinians, and conversely making Palestinians aware of the need to talk to and eventually reach a deal with Israel.

At present, Gush Shalom activists are mainly involved in daily struggle at Palestinian villages that have their land confiscated by the West Bank barrier, erected to stop suicide bombers. Gush activists are to be found, together with those of other Israeli movements like Ta'ayush and Anarchists Against the Wall, joining the Palestinian villagers of Bil'in in the weekly non-violent protest marches held to protest confiscation of more than half of the village lands.

Although Gush Shalom earned itself respect among peace-seeking Israelis as well as in the United States and Europe, it is regarded by mainstream Israelis as a purely pro-Palestinian movement.

United Kingdom

From 1934 the Peace Pledge Union gained many adherents to its pledge, "I renounce war and will never support or sanction another." Its support diminished considerably with the outbreak of war in 1939, but it remained the focus of pacifism in the post-war years.

Post–World War II peace-movement efforts in the United Kingdom were initially focused on the dissolution of the British Empire and the rejection of imperialism by the United States and Union of Soviet Socialist Republics. The anti-nuclear movement sought to "opt out" of the Cold War and rejected such ideas as "Britain's Little Independent Nuclear Deterrent" in part on the grounds that it (BLIND) was in contradiction even with MAD.

The VSC (Vietnam Solidarity Campaign) led by Tariq Ali mounted several very large and violent demonstrations against the Vietnam war in 67/68 but the first anti Vietnam demonstration was at the American Embassy in London and took place in 1965.

StWC Placard

The peace movement was later associated with the Peace camp movement as Labour moved "more to the centre" under Prime Minister Tony Blair. By early 2003, the peace and anti-war movement, mostly grouped together under the banner of the Stop the War Coalition, was powerful enough to cause several of Blair's cabinet to resign, and hundreds of Labour Party MPs to vote against their government. Blair's motion to support militarily the U.S. plan to invade Iraq continued only due to support from the UK Conservative Party. Protests against the invasion of Iraq were particularly vocal in Britain. Polls suggested that without UN Security Council approval, the UK public was very much opposed to involvement, and over two million people protested in Hyde Park (the previous largest demonstration in the UK having had around 600,000).

The peace movement has seen 'pop-up' newspapers, pirate radio stations and plays over the years. The UK had its first permanent national press newspaper with strong and consistent anti-war, pro peace movement themes in 2012. The presence of national media backing of the peace movement helped the newspaper grow to a monthly readership of almost 4 million without any financial investment or backing by 2015.

The primary function of the National Peace Congress was to provide opportunities for consultation and joint activities between its affiliated members, to help create an informed public opinion on the issues of the day and to convey to the government of the day the views of the substantial section of British life represented by its affiliated membership. The NPC folded in 2000 to be replaced in 2001 by Network for Peace, which was set up to continue the networking role of NPC.

United States

During the tail end of the Cold War, U.S. peace activists largely concentrated on slowing the superpower arms race in the belief that this would reduce the possibility of nuclear war between the U.S. and the USSR. As the Reagan administration accelerated military spending and adopted a tough, challenging stance to the Russians, peace groups such as the Nuclear Freeze and Beyond War sought to educate the public on what they believed was the inherent riskiness and ruinous cost of this policy. Outreach to individual citizens in the Soviet Union and mass meetings, using then-new satellite link technology, were part of peacemaking activities in the 1980s. In 1981, Thomas started the longest uninterrupted peace vigil in U.S. history. He was later joined at Lafayette Square by anti-nuclear activists Concepcion Picciotto and Ellen Thomas.

In response to Iraq's invasion of Kuwait in 1990, President George H. W. Bush began preparations for a mideast war. Peace activists were starting to find their groove just before the Gulf War was launched in February 1991, with well-attended rallies, especially on the west coast. However, the ground war was over in less than a week. A lopsided Allied victory and a media-incited wave of patriotic sentiment washed over the protest movement before it could develop traction.

During the 1990s, peacemakers' priorities included seeking a solution to the Israeli–Palestinian impasse, belated efforts at humanitarian assistance to war-torn regions such as Bosnia and Rwanda, and Iraq; American peace activists brought medicine into Iraq in defiance of U.S. law, in some cases enduring heavy fines and imprisonment in retaliation. Some of the principal groups involved were Voices in the Wilderness and the Fellowship of Reconciliation.

Protesters against the Iraq War in New York.

Before, during, and after the War in Iraq began, a concerted protest effort began in the United States. On February 15, 2003 a series of protests across the globe took place with events in approximately 800 cities. In March 2003, just before the U.S. and British Military led invasion of Iraq, a protest mobilization called "The World Says No to War" led to as many as 500,000 protestors in cities across the U.S. However, many protest organizations have persisted as the United States has maintained a military and corporate presence in Iraq.

U.S. activist groups including United for Peace and Justice, Code Pink (Women Say No To War), Iraq Veterans Against the War, Military Families Speak Out (MFSO), Not In Our Name, A.N.S.W.E.R., Veterans for Peace, and The World Can't Wait continued to protest against the Iraq War. Methods of protest include rallies and marches, impeachment petitions, the staging of a war-crimes tribunal in New York (to investigate crimes and alleged abuses of power of the Bush administration), bringing Iraqi women to tour the U.S. and tell their side of the story, street theater and independent filmmaking, high-profile appearances by anti-war activists such as Scott Ritter, Janis Karpinski, and Dahr Jamail, resisting military recruiting on college campuses, withholding tax monies, mass letter-writing to legislators and newspapers, blogging, music, and guerrilla theater. Independent media producers continue to broadcast, podcast and Web-host programs about the movement against the Iraq War.

Starting in 2005, opposition to military action against Iran started in the United States, the United Kingdom and elsewhere, including the creation of the Campaign Against Sanctions and Military Intervention in Iran. By August 2007, fears of an imminent United States and/or Israeli attack on Iran had increased to the level that several Nobel Prize winners, Shirin Ebadi (Nobel Peace Prize 2003), Mairead Corrigan-Maguire and Betty Williams (joint Nobel Peace Prize 1976), Harold Pinter (Nobel Prize for Literature 2005) and Jody Williams (Nobel Peace Prize 1997), along with several anti-war groups, including The Israeli Committee for a Middle East Free from Atomic, Biological and Chemical Weapons, Campaign for Nuclear Disarmament, CASMII, Code Pink and many others, warned about what they believed was the imminent risk of a "war of an unprecedented scale, this time against Iran", especially expressing concern that an attack on Iran using nuclear weapons had "not been ruled out". They called for "the dispute about Iran's nuclear program, to be resolved through peaceful means" and a call for Israel, "as the only Middle Eastern state suspected of possession of nuclear weapons", to join the Nuclear Non-Proliferation Treaty.

Social Justice

College students live in tents for a week to heighten awareness of what they believe is the need for social justice

Social justice is the fair and just relation between the individual and society. This is measured by the explicit and tacit terms for the distribution of wealth, opportunities for personal activity and social privileges. In Western as well as in older Asian cultures, the concept of social justice has often referred to the process of ensuring that individuals fulfill their societal roles and receive what was their due from society. In the current global grassroots movements for social justice, the emphasis has been on the breaking of barriers for social mobility, the creation of safety nets and economic justice.

Social justice assigns rights and duties in the institutions of society, which enables people to receive the basic benefits and burdens of cooperation. The relevant institutions often include taxation, social insurance, public health, public school, public services, labour law and regulation of markets, to ensure fair distribution of wealth, and equal opportunity.

Interpretations that relate justice to a reciprocal relationship to society are mediated by differences in cultural traditions, some of which emphasize the individual responsibility toward society and others the equilibrium between access to power and its responsible use. Hence, social justice is invoked today while reinterpreting historical figures such as Bartolomé de las Casas, in philosophical debates about differences among human beings, in efforts for gender, racial and social equality, for advocating justice for migrants, prisoners, the environment, and the physically and mentally disabled.

While the concept of social justice can be traced through the theology of Augustine of Hippo and the philosophy of Thomas Paine, the term "social justice" became used explicitly from the 1840s. A Jesuit priest named Luigi Taparelli is typically credited with coining the term, and it spread during the revolutions of 1848 with the work of Antonio Rosmini-Serbati. In the late industrial revolution, progressive American legal scholars began to use the term more, particularly Louis Brandeis and Roscoe Pound. From the early 20th century it was also embedded in international law and institutions; the preamble to establish the International Labour Organization recalled that "universal and lasting peace can be established only if it is based upon social justice." In the later 20th century, social justice was made central to the philosophy of the social contract, primarily by John Rawls in *A Theory of Justice* (1971). In 1993, the Vienna Declaration and Programme of Action treats social justice as a purpose of human rights education.

History

The different concepts of justice, as discussed in ancient Western philosophy, were typically centered upon the community.

Plato

- Plato wrote in The Republic that it would be an ideal state that "every member of the community must be assigned to the class for which he finds himself best fitted." In an article for J.N.V University, author D.R. Bhandari says, "Justice is, for Plato, at once a part of human virtue and the bond, which joins man together in society. It is the identical quality that makes good and social. Justice is an order and duty of the parts of the soul, it is to the soul as health is to the body. Plato says that justice is not mere strength, but it is a harmonious strength. Justice is not the right of the stronger but the effective harmony of the whole. All moral conceptions revolve about the good of the whole-individual as well as social".
- Aristotle believed rights existed only between free people, and the law should take "account in the first instance of relations of inequality in which individuals are treated in proportion to their worth and only secondarily of relations of equality." Reflecting this time when slavery and subjugation of women was typical, ancient views of justice tended to reflect the

rigid class systems that still prevailed. On the other hand, for the privileged groups, strong concepts of fairness and the community existed. Distributive justice was said by Aristotle to require that people were distributed goods and assets according to their merit.

Aristotle

Socrates

- Socrates (through Plato's dialogue Crito) is attributed with developing the idea of a social contract, whereby people ought to follow the rules of a society, and accept its burdens because they have accepted its benefits. During the Middle Ages, religious scholars particularly, such as Thomas Aquinas continued discussion of justice in various ways, but ultimately connected being a good citizen to the purpose of serving God.

After the Renaissance and Reformation, the modern concept of social justice, as developing human potential, began to emerge through the work of a series of authors. Baruch Spinoza in *On the Improvement of the Understanding* (1677) contended that the one true aim of life should be to acquire "a human character much more stable than [one's] own", and to achieve this "pitch of perfection... The chief good is that he should arrive, together with other individuals if possible, at the possession of the aforesaid character." During the enlightenment and responding to the French and American Revolutions, Thomas Paine similarly wrote in *The Rights of Man* (1792) society should give "genius a fair and universal chance" and so "the construction of government ought to be such as to bring forward... all that extent of capacity which never fails to appear in revolutions."

"Social justice" was coined by Jesuit priest Luigi Taparelli in the 1840s.

The first modern usage of the specific term "social justice" is typically attributed to Catholic thinkers from the 1840s, including the Jesuit Luigi Taparelli in *Civiltà Cattolica*, based on the work of St. Thomas Aquinas. He argued that rival capitalist and socialist theories, based on subjective Cartesian thinking, undermined the unity of society present in Thomistic metaphysics as neither were sufficiently concerned with moral philosophy. Writing in 1861, the influential British philosopher and economist, John Stuart Mill stated in *Utilitarianism* his view that "Society should treat all equally well who have deserved equally well of it, that is, who have deserved equally well absolutely. This is the highest abstract standard of social and distributive justice; towards which all institutions, and the efforts of all virtuous citizens, should be made in the utmost degree to converge."

In the later 19th and early 20th century, social justice became an important theme in American political and legal philosophy, particularly in the work of John Dewey, Roscoe Pound and Louis Brandeis. One of the prime concerns was the *Lochner era* decisions of the US Supreme Court to strike down legislation passed by state governments and the Federal government for social and economic improvement, such as the eight-hour day or the right to join a trade union. After the First World War, the founding document of the International Labour Organization took up the same terminology in its preamble, stating that "peace can be established only if it is based on social justice". From this point, the discussion of social justice entered into mainstream legal and academic discourse. In the late 20th century, a number of liberal and conservative thinkers, notably

Friedrich von Hayek rejected the concept by stating that it did not mean anything, or meant too many things. However the concept remained highly influential, particularly with its promotion by philosophers such as John Rawls.

Contemporary Theory

Cosmic Values

Hunter Lewis' work promoting natural healthcare and sustainable economies advocates for conservation as a key premise in social justice. His manifesto on sustainability ties the continued thriving of human life to real conditions, the environment supporting that life, and associates injustice with the detrimental effects of unintended consequences of human actions. Quoting classical Greek thinkers like Epicurus on the good of pursuing happiness, Hunter also cites ornithologist, naturalist, and philosopher Alexander Skutch in his book Moral Foundations:

The common feature which unites the activities most consistently forbidden by the moral codes of civilized peoples is that by their very nature they cannot be both habitual and enduring, because they tend to destroy the conditions which make them possible.

Pope Benedict XVI cites Teilhard de Chardin in a vision of the cosmos as a 'living host' embracing an understanding of ecology that includes humanity's relationship to others, that pollution affects not just the natural world but interpersonal relations as well. Cosmic harmony, justice and peace are closely interrelated:

If you want to cultivate peace, protect creation.

John Rawls

Political philosopher John Rawls draws on the utilitarian insights of Bentham and Mill, the social contract ideas of John Locke, and the categorical imperative ideas of Kant. His first statement of principle was made in *A Theory of Justice* where he proposed that, "Each person possesses an inviolability founded on justice that even the welfare of society as a whole cannot override. For this reason justice denies that the loss of freedom for some is made right by a greater good shared by others." A deontological proposition that echoes Kant in framing the moral good of justice in absolutist terms. His views are definitively restated in *Political Liberalism* where society is seen "as a fair system of co-operation over time, from one generation to the next".

All societies have a basic structure of social, economic, and political institutions, both formal and informal. In testing how well these elements fit and work together, Rawls based a key test of legitimacy on the theories of social contract. To determine whether any particular system of collectively enforced social arrangements is legitimate, he argued that one must look for agreement by the people who are subject to it, but not necessarily to an objective notion of justice based on coherent ideological grounding. Obviously, not every citizen can be asked to participate in a poll to determine his or her consent to every proposal in which some degree of coercion is involved, so one has to assume that all citizens are reasonable. Rawls constructed an argument for a two-stage process to determine a citizen's hypothetical agreement:

- The citizen agrees to be represented by X for certain purposes, and, to that extent, X holds these powers as a trustee for the citizen.
- X agrees that enforcement in a particular social context is legitimate. The citizen, therefore, is bound by this decision because it is the function of the trustee to represent the citizen in this way.

This applies to one person who represents a small group (e.g., the organiser of a social event setting a dress code) as equally as it does to national governments, which are ultimate trustees, holding representative powers for the benefit of all citizens within their territorial boundaries. Governments that fail to provide for welfare of their citizens according to the principles of justice are not legitimate. To emphasise the general principle that justice should rise from the people and not be dictated by the law-making powers of governments, Rawls asserted that, "There is a general presumption against imposing legal and other restrictions on conduct without sufficient reason. But this presumption creates no special priority for any particular liberty." This is support for an unranked set of liberties that reasonable citizens in all states should respect and uphold — to some extent, the list proposed by Rawls matches the normative human rights that have international recognition and direct enforcement in some nation states where the citizens need encouragement to act in a way that fixes a greater degree of equality of outcome. According to Rawls, the basic liberties that every good society should guarantee are,

- Freedom of thought;
- Liberty of conscience as it affects social relationships on the grounds of religion, philosophy, and morality;
- Political liberties (e.g., representative democratic institutions, freedom of speech and the press, and freedom of assembly);
- Freedom of association;
- Freedoms necessary for the liberty and integrity of the person (namely: freedom from slavery, freedom of movement and a reasonable degree of freedom to choose one's occupation); and
- Rights and liberties covered by the rule of law.

Thomas Pogge

Thomas Pogge's arguments pertain to a standard of social justice that creates human rights deficits. He assigns responsibility to those who actively cooperate in designing or imposing the social institution, that the order is foreseeable as harming the global poor and is reasonably avoidable. Pogge argues that social institutions have a negative duty - not harm the poor.

Pogge speaks of "institutional cosmopolitanism" and assigns responsibility to institutional schemes for deficits of human rights. An example given is slavery and third parties. A third party should not recognize or enforce slavery. The institutional order should be held responsible only for deprivations of human rights that it establishes or authorizes. The current institutional design, he says, systematically harms developing economies by enabling corporate tax evasion, illicit financial flows, corruption, trafficking of people and weapons. Joshua Cohen disputes his claims based on the fact that some poor countries have done well with the current institutional design. Elizabeth Kahn argues that some of these responsibilities should apply globally.

United Nations

The United Nations' 2006 document *Social Justice in an Open World: The Role of the United Nations*, states that "Social justice may be broadly understood as the fair and compassionate distribution of the fruits of economic growth"

The term "social justice" was seen by the U.N. "as a substitute for the protection of human rights [and] first appeared in United Nations texts during the second half of the 1960s. At the initiative of the Soviet Union, and with the support of developing countries, the term was used in the Declaration on Social Progress and Development, adopted in 1969."

The same document reports, "From the comprehensive global perspective shaped by the United Nations Charter and the Universal Declaration of Human Rights, neglect of the pursuit of social justice in all its dimensions translates into de facto acceptance of a future marred by violence, repression and chaos." The report concludes, "Social justice is not possible without strong and coherent redistributive policies conceived and implemented by public agencies."

The same UN document offers a concise history: "The notion of social justice is relatively new. None of history's great philosophers—not Plato or Aristotle, or Confucius or Averroes, or even Rousseau or Kant—saw the need to consider justice or the redress of injustices from a social perspective. The concept first surfaced in Western thought and political language in the wake of the industrial revolution and the parallel development of the socialist doctrine. It emerged as an expression of protest against what was perceived as the capitalist exploitation of labour and as a focal point for the development of measures to improve the human condition. It was born as a revolutionary slogan embodying the ideals of progress and fraternity. Following the revolutions that shook Europe in the mid-1800s, social justice became a rallying cry for progressive thinkers and political activists. By the mid-twentieth century, the concept of social justice had become central to the ideologies and programmes of virtually all the leftist and centrist political parties around the world"

Religious Perspectives

Hinduism

The present-day Jāti hierarchy is undergoing changes for a variety of reasons including 'social justice', which is a politically popular stance in democratic India. Institutionalized affirmative action has promoted this. The disparity and wide inequalities in social behaviour of the jātis – exclusive, endogamous communities centred on traditional occupations – has led to various reform movements in Hinduism. While legally outlawed, the caste system remains strong in practice.

Islam

The Quran contains numerous references to elements of social justice. For example, one of

Islam's Five Pillars is Zakāt, or alms-giving. Charity and assistance to the poor – concepts central to social justice – are and have historically been important parts of the Islamic faith.

In Muslim history, Islamic governance has often been associated with social justice. Establishment of social justice was one of the motivating factors of the Abbasid revolt against the Umayyads. The Shi'a believe that the return of the *Mahdi* will herald in "the messianic age of justice" and the Mahdi along with the Isa (Jesus) will end plunder, torture, oppression and discrimination.

For the Muslim Brotherhood the implementation of social justice would require the rejection of consumerism and communism. The Brotherhood strongly affirmed the right to private property as well as differences in personal wealth due to factors such as hard work. However, the Brotherhood held Muslims had an obligation to assist those Muslims in need. It held that *zakat* (alms-giving) was not voluntary charity, but rather the poor had the right to assistance from the more fortunate. Most Islamic governments therefore enforce the *zakat* through taxes.

Though monetary donations are the most practiced way of zakat, Islam is deeply rooted in the tenets of volunteerism and social activism. Areas of one's communities which require assistance and beneficiaries must be a Muslim's foci if need be, rather than strictly her or his personal or superficial wants. For example, the ecological well-being of the planet (i.e.: animal rights, global warming, natural resources degradation); locally, nationally, globally, is a campaign to which every Muslim must adhere. Many Muslims practice this today by ensuring that they produce minimal waste, give to charity what they no longer need, and spend time in prayer and meditation upon the bounties of nature so as to more mindfully approach all that is provided by nature, and ultimately, Allah. Other areas of society in need may be the safety and security of minority populations, i.e.: women or persons of color, children, the elderly, the developmentally or physically disabled, animals, et al.

Judaism

In *To Heal a Fractured World: The Ethics of Responsibility*, Rabbi Jonathan Sacks states that social justice has a central place in Judaism. One of Judaism's most distinctive and challenging ideas is its ethics of responsibility reflected in the concepts of simcha ("gladness" or "joy"), tzedakah ("the religious obligation to perform charity and philanthropic acts"), chesed ("deeds of kindness"), and tikkun olam ("repairing the world").

Christianity

Methodism

From its founding, Methodism was a Christian social justice movement. Under John Wesley's direction, Methodists became leaders in many social justice issues of the day, including the prison reform and abolition movements. Wesley himself was among the first to preach for slaves rights attracting significant opposition.

Today, social justice plays a major role in the United Methodist Church. The *Book of Discipline of the United Methodist Church* says, "We hold governments responsible for the protection of the rights of the people to free and fair elections and to the freedoms of speech, religion, assembly, communications media, and petition for redress of grievances without fear of reprisal; to the right to privacy; and to the guarantee of the rights to adequate food, clothing, shelter, education, and health care." The United Methodist Church also teaches population control as part of its doctrine.

Catholicism

Catholic social teaching consists of those aspects of Roman Catholic doctrine which relate to matters dealing with the respect of the individual human life. A distinctive feature of Catholic social doctrine is its concern for the poorest and most vulnerable members of society. Two of the seven key areas of "Catholic social teaching" are pertinent to social justice:

- Life and dignity of the human person: The foundational principle of all "Catholic Social Teaching" is the sanctity of all human life and the inherent dignity of every human person, from conception to natural death. Human life must be valued above all material possessions.
- Preferential option for the poor and vulnerable: Catholics believe Jesus taught that on the Day of Judgement God will ask what each person did to help the poor and needy: "Amen, I say to you, whatever you did for one of these least brothers of mine, you did for me." The Catholic Church believes that through words, prayers and deeds one must show solidarity with, and compassion for, the poor. The moral test of any society is "how it treats its most vulnerable members. The poor have the most urgent moral claim on the conscience of the nation. People are called to look at public policy decisions in terms of how they affect the poor."

Even before it was propounded in the Catholic social doctrine, social justice appeared regularly in the history of the Catholic Church:

- Pope Leo XIII, who studied under Taparelli, published in 1891 the encyclical *Rerum novarum* (On the Condition of the Working Classes; lit. "On new things"), rejecting both socialism and capitalism, while defending labor unions and private property. He stated that society should be based on cooperation and not class conflict and competition. In this document, Leo set out the Catholic Church's response to the social instability and labor conflict that had arisen in the wake of industrialization and had led to the rise of socialism. The Pope advocated that the role of the State was to promote social justice through the protection of rights, while the Church must speak out on social issues in order to teach correct social principles and ensure class harmony.
- The encyclical *Quadragesimo anno* (On Reconstruction of the Social Order, literally "in the fortieth year") of 1931 by Pope Pius XI, encourages a living wage, subsidiarity, and advocates that social justice is a personal virtue as well as an attribute of the social order, saying that society can be just only if individuals and institutions are just.
- Pope John Paul II added much to the corpus of the Catholic social teaching, penning three encyclicals which focus on issues such as economics, politics, geo-political situations, ownership of the means of production, private property and the "social mortgage", and private property. The encyclicals *Laborem exercens, Sollicitudo rei socialis*, and *Centesimus annus* are just a small portion of his overall contribution to Catholic social justice. Pope John Paul II was a strong advocate of justice and human rights, and spoke forcefully for the poor. He addresses issues such as the problems that technology can present should it

be misused, and admits a fear that the "progress" of the world is not true progress at all, if it should denigrate the value of the human person. He argued in *Centesimus annus* that private property, markets, and honest labor were the keys to alleviating the miseries of the poor and to enabling a life that can express the fullness of the human person.

- Pope Benedict XVI's encyclical *Deus caritas est* ("God is Love") of 2006 claims that justice is the defining concern of the state and the central concern of politics, and not of the church, which has charity as its central social concern. It said that the laity has the specific responsibility of pursuing social justice in civil society and that the church's active role in social justice should be to inform the debate, using reason and natural law, and also by providing moral and spiritual formation for those involved in politics.
- The official Catholic doctrine on social justice can be found in the book *Compendium of the Social Doctrine of the Church*, published in 2004 and updated in 2006, by the Pontifical Council *Iustitia et Pax*.

The Catechism (1928–1948) contain more detail of the Church's view of social justice.

Criticism

Many authors criticize the idea that there exists an objective standard of social justice. Moral relativists deny that there is any kind of objective standard for justice in general. Non-cognitivists, moral skeptics, moral nihilists, and most logical positivists deny the epistemic possibility of objective notions of justice. Political realists believe that any ideal of social justice is ultimately a mere justification for the status quo.

Many other people accept some of the basic principles of social justice, such as the idea that all human beings have a basic level of value, but disagree with the elaborate conclusions that may or may not follow from this. One example is the statement by H. G. Wells that all people are "equally entitled to the respect of their fellowmen."

On the other hand, some scholars reject the very idea of social justice as meaningless, religious, self-contradictory, and ideological, believing that to realize any degree of social justice is unfeasible, and that the attempt to do so must destroy all liberty. Perhaps the most complete rejection of the concept of social justice comes from Friedrich Hayek of the Austrian School of economics:

There can be no test by which we can discover what is 'socially unjust' because there is no subject by which such an injustice can be committed, and there are no rules of individual conduct the observance of which in the market order would secure to the individuals and groups the position which as such (as distinguished from the procedure by which it is determined) would appear just to us. [Social justice] does not belong to the category of error but to that of nonsense, like the term 'a moral stone'.

Ben O'Neill of the University of New South Wales argues that, for proponents of "social justice":

the notion of "rights" is a mere term of entitlement, indicative of a claim for any possible desirable good, no matter how important or trivial, abstract or tangible, recent or ancient. It is merely an assertion of desire, and a declaration of intention to use the language of rights to acquire said desire.

In fact, since the program of social justice inevitably involves claims for government provision of goods, paid for through the efforts of others, the term actually refers to an intention to use *force* to acquire one's desires. Not to earn desirable goods by rational thought and action, production and voluntary exchange, but to go in there and forcibly take goods from those who can supply them!

Janusz Korwin-Mikke states, "Either 'social justice' has the same meaning as 'justice' – or not. If so – why use the additional word 'social?' We lose time, we destroy trees to obtain paper necessary to print this word. If not, if 'social justice' means something different from 'justice' – then 'something different from justice' is by definition 'injustice.'"

Sociologist Carl L. Bankston has argued that a secular, leftist view of social justice entails viewing the redistribution of goods and resources as based on the rights of disadvantaged categories of people, rather than on compassion or national interest. Bankston maintains that this secular version of social justice became widely accepted due to the rise of demand-side economics and to the moral influence of the civil rights movement.

Social Justice Movements

Social justice is also a concept that is used to describe the movement towards a socially just world, e.g., the Global Justice Movement. In this context, social justice is based on the concepts of human rights and equality, and can be defined as *"the way in which human rights are manifested in the everyday lives of people at every level of society"*.

A number of movements are working to achieve social justice in society. These movements are working towards the realization of a world where all members of a society, regardless of background or procedural justice, have basic human rights and equal access to the benefits of their society.

Liberation Theology

Liberation theology is a movement in Christian theology which conveys the teachings of Jesus Christ in terms of a liberation from unjust economic, political, or social conditions. It has been described by proponents as "an interpretation of Christian faith through the poor's suffering, their struggle and hope, and a critique of society and the Catholic faith and Christianity through the eyes of the poor", and by detractors as Christianity perverted by Marxism and Communism.

Although liberation theology has grown into an international and inter-denominational movement, it began as a movement within the Catholic Church in Latin America in the 1950s–1960s. It arose principally as a moral reaction to the poverty caused by social injustice in that region. It achieved prominence in the 1970s and 1980s. The term was coined by the Peruvian priest, Gustavo Gutiérrez, who wrote one of the movement's most famous books, *A Theology of Liberation* (1971). According to Sarah Kleeb, "Marx would surely take issue," she writes, "with the appropriation of his works in a religious context... there is no way to reconcile Marx's views of religion with those of Gutierrez, they are simply incompatible. Despite this, in terms of their understanding of the necessity of a just and righteous world, and the nearly inevitable obstructions along such a path, the two have much in common; and, particularly in the first edition of [A Theology of Liberation], the use of Marxian theory is quite evident."

Other noted exponents are Leonardo Boff of Brazil, Carlos Mugica of Argentina, Jon Sobrino of El Salvador, and Juan Luis Segundo of Uruguay.

Health Care

Social justice has more recently made its way into the field of bioethics. Discussion involves topics such as affordable access to health care, especially for low income households and families. The discussion also raises questions such as whether society should bear healthcare costs for low income families, and whether the global marketplace is the best way to distribute healthcare. Ruth Faden of the Johns Hopkins Berman Institute of Bioethics and Madison Powers of Georgetown University focus their analysis of social justice on which inequalities matter the most. They develop a social justice theory that answers some of these questions in concrete settings.

Social injustices occur when there is a preventable difference in health states among a population of people. These social injustices take the form of health inequities when negative health states such as malnourishment, and infectious diseases are more prevalent in impoverished nations. These negative health states can often be prevented by providing social and economic structures such as primary healthcare which ensures the general population has equal access to health care services regardless of income level, gender, education or any other stratifying factors. Integrating social justice with health inherently reflects the social determinants of health model without discounting the role of the bio-medical model.

Human Rights Education

The Vienna Declaration and Programme of Action affirm that "Human rights education should include peace, democracy, development and social justice, as set forth in international and regional human rights instruments, in order to achieve common understanding and awareness with a view to strengthening universal commitment to human rights."

Social Equality

Social equality is a state of affairs in which all people within a specific society or isolated group have the same status in certain respects, including civil rights, freedom of speech, property rights and equal access to certain social goods and services. However, it also includes concepts of health equity, economic equality and other social securities. It also includes equal opportunities and obligations, and so involves the whole of society. Social equality requires the absence of legally enforced social class or caste boundaries and the absence of discrimination motivated by an inalienable part of a person's identity. For example, sex, gender, race, age, sexual orientation, origin, caste or class, income or property, language, religion, convictions, opinions, health or disability must not result in unequal treatment under the law and should not reduce opportunities unjustifiably.

"Equal opportunities" is interpreted as being judged by ability, which is compatible with a free-market economy. Relevant problems are horizontal inequality – the inequality of two persons of *same* origin and ability and differing opportunities given to individuals – such as in education or by inherited capital.

Conceivements of social equality may vary per philosophy and individual and other than egalitarianism it does not necessarily require all social inequalities to be eliminated by artificial means but instead often recognizes and respects natural differences between people.

Ontological

The standard of equality that states everyone is created equal at birth is called ontological equality. This type of equality can be seen in many different places like the Declaration of Independence. This early document, which states many of the values of the United States of America, has this idea of equality embedded in it. It clearly states that "all men are created equal, that they are endowed by their Creator with certain unalienable Rights". The statement reflects the philosophy of John Locke and his idea that we are all equal in certain natural rights. Although this standard of equality is seen in documents as important as the Declaration of Independence, it is "one not often invoked in policy debates these days". However this notion of equality is often used to justify inequalities such as material inequality. Dalton Conley claims that ontological equality is used to justify material inequality by putting a spotlight on the fact, legitimated by theology, that "the distribution of power and resources here on earth does not matter, because all of us are equally children of God and will have to face our maker upon dying". Dalton Conley, the author of *You May Ask Yourself*, claims that ontological equality can also be used to put forth the notion that poverty is virtue. Luciano Floridi, author of a book about information, wrote about what he calls the ontological equality principle. His work on information ethics raises the importance of equality when presenting information. Here is a short sample of his work:

> Information ethics is impartial and universal because it brings to ultimate completion the process of enlargement of the concept of what may count as a centre of a (no matter how minimal) moral claim, which now includes every instance of being understood informationally, no matter whether physically implemented or not. In this respect information ethics holds that every entity as an expression of being, has a dignity constituted by its mode of existence and essence (the collection of all the elementary properties that constitute it for what it is), which deserve to be respected (at least in a minimal and overridable sense), and hence place moral claims on the interacting agent and ought to contribute to the constraint and guidance of his ethical decisions and behaviour.

Floridi goes onto claim that this "ontological equality principle means that any form of reality (any instance of information/being), simply for the fact of being what it is, enjoys a minimal, initial, overridable, equal right to exist and develop in a way which is appropriate to its nature." Values in his claims correlate to those shown in the sociological textbook *You May Ask Yourself* by Dalton Conley. The notion of "ontological equality" describes equality by saying everything is equal by nature. Everyone is created equal at birth. Everything has equal right to exist and develop by its nature.

Opportunity

Another standard of equality is equality of opportunity, "the idea that everyone has an equal chance to achieve wealth, social prestige, and power because the rules of the game, so to speak, are the same for everyone". This concept can be applied to society by saying that no one has a head start. This means that, for any social equality issue dealing with wealth, social prestige, power, or any of that sort, the equality of opportunity standard can defend the idea that everyone had the same start. This views society almost as a game and any of the differences in equality are due to luck and playing the "game" to one's best ability. Conley gives an example of this standard of equality by using a game of Monopoly to describe society. He claims that "Monopoly follows the rules of equality of opportunity" by explaining that everyone had an equal chance when starting the game and any differences were a result of the luck of the dice roll and the skill of the player to make choices to benefit their wealth. Comparing this example to society, the standard of equality of opportunity eliminates inequality because the rules of the games in society are still fair and the same for all; therefore making any existing inequalities in society fair. Lesley A. Jacobs, the author of *Pursuing Equal Opportunities: The Theory and Practice of Egalitarian Justice*, talks about equality of opportunity and its importance relating to egalitarian justice. Jacobs states that

> at the core of equality of opportunity... is the concept that in competitive procedures designed for the allocation of scarce resources and the distribution of the benefits and burdens of social life, those procedures should be governed by criteria that are relevant to the particular goods at stake in the competition and not by irrelevant considerations such as race, religion, class, gender, disability, sexual orientation, ethnicity, or other factors that may hinder some of the competitors' opportunities at success. (Jacobs, 10).

This concept points out factors like race, gender, class etc. that should not be considered when talking about equality through this notion. Conley also mentions that this standard of equality is at the heart of a bourgeois society, such as a modern capitalist society, or "a society of commerce in which the maximization of profit is the primary business incentive". It was the equal opportunity ideology that civil rights activists adopted in the era of the Civil Rights Movement in the 1960s. This ideology was used by them to argue that Jim Crow laws were incompatible with the standard of equality of opportunity.

Condition

Another notion of equality introduced by Conley is equality of condition. Through this framework is the idea that everyone should have an equal starting point. Conley goes back to his example of a game of Monopoly to explain this standard. If the game of four started off with two players both having an advantage of $5,000 dollars to start off with and both already owning hotels and other property while the other two players both did not own any property and both started off with a $5,000 dollar deficit, then from a perspective of the standard of equality of condition, one can argue that the rules of the game "need to be altered in order to compensate for inequalities in the relative starting positions". From this we form policies in order to even equality which in result bring an efficient way to create fairer competition in society. Here is where social engineering comes into play where we change society in order to give an equality of condition to everyone based on race, gender, class, religion etc. when it is made justifiable that the proponents of the society makes it unfair for them.

Sharon E. Kahn, author of *Academic Freedom and the Inclusive University*, talks about equality of condition in their work as well and how it correlates to freedom of individuals. They claim that in order to have individual freedom there needs to be equality of condition "which requires much more than the elimination of legal barriers: it requires the creation of a level playing field that eliminates structural barriers to opportunity". Her work talks about the academic structure and its problem with equalities and claims that to "ensure equity...we need to recognize that the university structure and its organizational culture have traditionally privileged some and marginalized other;

we need to go beyond theoretical concepts of equality by eliminating systemic barriers that hinder the equal participation of members of all groups; we need to create and equality of condition, not merely an equality of opportunity". "Notions of equity, diversity, and inclusiveness begin with a set of premises about individualism, freedom and rights that take as given the existence of deeply rooted inequalities in social structure," therefore in order to have a culture of the inclusive university, it would have to "be based on values of equity; that is, equality of condition" eliminating all systemic barriers that go against equality.

Outcome

A fourth standard of equality is equality of outcome, which is "a position that argues each player must end up with the same amount regardless of the fairness". This ideology is predominately a Marxist philosophy that is concerned with equal distribution of power and resources rather than the rules of society. In this standard of equality the idea is that "everyone contributes to society and to the economy according to what they do best". Under this notion of equality, Conley states that "nobody will earn more power, prestige, and wealth by working harder".

When defining equality of outcome in education, "the goals should not be the liberal one of equality of access but equality of outcome for the median number of each identifiable non-educationally defined group, i.e. the average women, negro, or proletarian or rural dweller should have the same level of educational attainment as the average male, white, suburbanite". The outcome and the benefits from equality from education from this notion of equality promotes that all should have the same outcomes and benefits regardless of race, gender, religion etc. The equality of outcome in Hewitt's point of view is supposed to result in "a comparable range of achievements between a specific disadvantaged group – such as an ethnic minority, women, lone parents and the disabled – and society as a whole".

Equality of Outcome

In the satirical novel *Alice's Adventures in Wonderland*, the Dodo tells Alice "Everybody has won and all must have prizes", and one analyst suggested that this quote describes the controversial concept of *equality of outcome*.

Equality of outcome, equality of condition, or equality of results is a political concept which is central to some political ideologies and is used regularly in political discourse, often in contrast to the term equality of opportunity. It describes a state in which people have approximately the same material wealth and income, or in which the general economic conditions of their lives are similar. Achieving equal results generally entails reducing or eliminating material inequalities between individuals or households in a society, and usually involves a transfer of income or wealth from wealthier to poorer individuals, or adopting other measures to promote equality of condition. A related way of defining equality of outcome is to think of it as "equality in the central and valuable things in life." One account in the *Journal of Political Philosophy* suggested that the term meant, "equalising where people end up rather than where or how they begin," but described this sense of the term as "simplistic" since it failed to identify what was supposed to be made equal.

Comparisons with Related Concepts

Equality of outcome is often compared to related concepts of equality, particularly with *equality of opportunity*. Generally, most senses of the concept of *equality* are controversial, and are seen differently by people having different political perspectives, but of all of the terms relating to *equality*, *equality of outcome* is the most "controversial" or "contentious".

- Equality of opportunity. This conception generally describes fair competition for important jobs and positions such that contenders have equal chances to win such positions, and applicants are not judged or hampered by unfair or arbitrary discrimination. It entails the "elimination of arbitrary discrimination in the process of selection." The term is usually applied in workplace situations but has been applied in other areas as well such as housing, lending, and voting rights. The essence is that job seekers have "an equal chance to compete within the framework of goals and the structure of rules established," according to one view. It is generally seen as a procedural value of fair treatment by the rules.

  Equality of opportunity provides in a sense that all start the race of life at the same time. Equality of outcome attempts to ensure that everyone finishes at the same time.

  —*Mark Cooray, 1996*

- Equality of autonomy. This relatively new concept, a sort of hybrid notion, has been developed by philosopher Amartya Sen and can be thought of as "the ability and means to choose our life course should be spread as equally as possible across society." It is an equal shot at empowerment or a chance to develop up to his or her potential rather than equal goods or equal chances. In a teaching guide, *equality of autonomy* was explained as "equality in the degree of empowerment people have to make decisions affecting their lives, how much choice and control they have given their circumstances." Sen's approach requires "active intervention of institutions like the state into people's lives" but with an aim towards "fostering of people's self-creation rather than their living conditions." Sen argued that "the ability to convert incomes into opportunities is affected by a multiplicity of individual and social differences that mean some people will need more than others to achieve the same range of capabilities."

- Equality of process is related to the general notion of fair treatment, and can be thought of as "dealing with inequalities in treatment through discrimination by other individuals and groups, or by institutions and systems, including not being treated with dignity and respect," according to one definition.
- Equality of perception. This is an uncommonly used term meaning that "person should be perceived as being of equal worth."

Political Philosophy

In political philosophy, there are differing views whether equal outcomes are beneficial or not. One view is that there is a moral basis for equality of outcome, but that means to achieve such an outcome can be malevolent. *Equality of outcome* can be a good thing *after* it has been achieved since it reflects the natural "interdependence of citizens in a highly organized economy" and provides a "basis for social policies" which foster harmony and good will, including social cohesion and reduced jealousy. One writer suggested greater socioeconomic equality was "indispensable if we want to realise our shared commonsense values of societal fairness." Analyst Kenneth Cauthen in his 1987 book *The Passion for Equality* suggested that there were moral underpinnings for having equal outcomes because there is a common good––which people both contribute to and receive benefits from––and therefore should be enjoyed in common; Cauthen argued that this was a fundamental basis for both equality of opportunity as well as equality of outcome. Analyst George Packer, writing in the journal *Foreign Affairs*, argued that "inequality undermines democracy" in the United States partially because it "hardens society into a class system, imprisoning people in the circumstances of their birth." Packer elaborated that inequality "corrodes trust among fellow citizens" and compared it to an "odorless gas which pervades every corner" of the nation.

An opposing view is that equality of outcomes is not beneficial overall for society since it dampens motivation necessary for humans to achieve great things, such as new inventions, intellectual discoveries, and artistic breakthroughs. According to this view, economic wealth and social status are rewards needed to spur such activity, and with these rewards diminished, then achievements which will ultimately benefit everybody will not happen as frequently.

If equality of outcomes is seen as beneficial for society, and if people have differing levels of material wealth and social prestige in the present, then methods to transform a society towards one with greater equality of outcomes is problematic. A mainstream view is that mechanisms to achieve equal outcomes––to take a society and with unequal socioeconomic levels and force it to equal outcomes––are fraught with moral as well as practical problems since they often involve political coercion to compel the transfer.

And there is general agreement that outcomes matter. In one report in Britain, unequal outcomes in terms of personal wealth had a strong impact on average life expectancy, such that wealthier people tended to live seven years longer than poorer people, and that egalitarian nations tended to have fewer problems with societal issues such as mental illness, violence, teenage pregnancy, and other social problems. Authors of the book *The Spirit Level* contended that "more equal societies almost always do better" on other measures, and as a result, striving for equal outcomes can have overall beneficial effects for everybody.

Philosopher John Rawls, in his *A Theory of Justice* (1971), developed a "second principle of justice" that economic and social inequalities can only be justified if they benefit the most disadvantaged members of society. Further, Rawls claims that all economically and socially privileged positions must be open to all people equally. Rawls argues that the inequality between a doctor's salary and a grocery clerk's is only acceptable if this is the only way to encourage the training of sufficient numbers of doctors, preventing an unacceptable decline in the availability of medical care (which would therefore disadvantage everyone). Analyst Paul Krugman writing in *The New York Times* agreed with Rawls' position in which both equality of opportunity and equality of outcome were linked, and suggested that "we should try to create the society each of us would want if we didn't know in advance who we'd be." Krugman favored a society in which hard-working and talented people can get rewarded for their efforts but in which there was a "social safety net" created by taxes to help the less fortunate.

Economist Paul Krugman in 2008.

Conflation with Marxism, Socialism, and Communism

The German economist and philosopher Karl Marx is sometimes mistakenly characterized as an egalitarian and a proponent of equality of outcome, and the economic systems of socialism and communism are sometimes misconstrued as being based on equality of outcome. In reality Marx eschewed the entire concept of equality as abstract and bourgeois in nature, focusing his analysis on more concrete issues such as opposition to exploitation based on economic and materialist logic. Marx eschewed theorizing on moral concepts and refrained from advocating principles of justice. Marx's views on equality were informed by his analysis of the development of the productive forces in society.

Socialism is based on a principle of distribution whereby individuals receive compensation proportional to the amount of energy and labor they contribute to production ("To each according to his contribution"), which by definition precludes equal outcomes in income distribution.

In Marxist theory, communism is based on a principle whereby access to goods and services is based on free and open access (often referred to as distribution based on one's needs in the literature); Marx stressed free access to the articles of consumption. The "equality" in a communist society is thus not about total equality or equality of outcome, but about equal and free access to the articles of consumption. Marx argued that free access to consumption would enable individuals to overcome alienation.

Despite this, socialists, communists and Marxists believe that by eliminating exploitation, their respective principle of compensation will lead to greater equality than that found in capitalism because there would be no inequality arising from private ownership of productive property.

Comparing Equalities: Outcome vs Opportunity

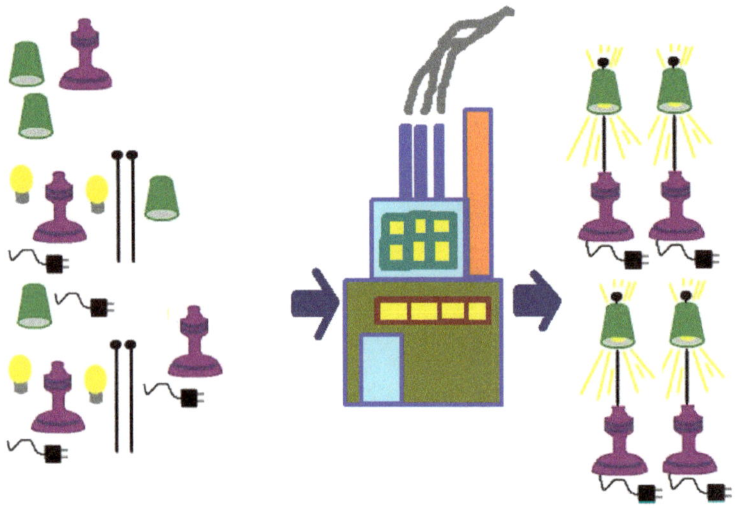

In a lamp assembly factory, for example, *equality of outcome* might mean that workers are all *paid equally* regardless of how many lamps of acceptable quality they make (which also implies that the workers cannot be sacked for producing too few lamps of acceptable quality). This can be contrasted with a payment system such as piece work, which requires that every worker is paid a fixed amount of money per lamp of acceptable quality that the worker makes.

Both *equality of outcome* and *equality of opportunity* have been contrasted to a great extent. When evaluated in a simple context, the more preferred term in contemporary political discourse is *equality of opportunity* (or, meaning the same thing, the common variant "equal opportunities") which the public, as well as individual commentators, see as the nicer or more "well-mannered" of the two terms. And the term *equality of outcome* is seen as more controversial and is viewed skeptically. A mainstream political view is that the comparison of the two terms is valid, but that they are somewhat mutually exclusive in the sense that striving for either type of equality would require sacrificing the other to an extent, and that achieving *equality of opportunity* necessarily brings about "certain inequalities of outcome." For example, striving for equal outcomes might require discriminating between groups to achieve these outcomes; or striving for equal opportunities in some types of treatment might lead to unequal results. Policies that seek an equality of outcome often require a deviation from the strict application of concepts such as meritocracy, and legal notions of equality before the law for all citizens. 'Equality seeking' policies may also have a redistributive focus.

The two concepts, however, are not always cleanly contrasted, since the notion of *equality* is complex. Some analysts see the two concepts not as polar opposites but as highly related such that they can not be understood without considering the other term. One writer suggested it was unrealistic to think about *equality of opportunity* in isolation, without considering inequalities of income and wealth. Another agreed that it is impossible to understand *equality* without some assessment of outcomes. A third writer suggested that trying to pretend that the two concepts were "fundamentally different" was an error along the lines of a conceit.

Philosopher Friedrich Nietzsche in 1875.

In contemporary political discourse, of the two concepts, *equality of outcome* has sometimes been criticized as the "politics of envy" and is often seen as more "controversial" than *equality of opportunity*. One wrote that "equality of opportunity is then set up as the mild-mannered alternative to the craziness of outcome equality." One theorist suggested that an over-emphasis on either type of equality can "come into conflict with individual freedom and merit." The philosopher Friedrich Nietzsche did not like either type of *equality* and was opposed in principle to democracy. He associated all equality with what he termed "slave morality."

Critics of equality of opportunity note that while it is relatively easier to deal with unfairness for people with different races or genders, it is much harder to deal with social class since "one can never entirely extract people from their ancestry and upbringing." As a result, critics contend that efforts to bring fairness by equal opportunity are stymied by the difficulty of people having differing starting points at the beginning of the socio-economic competition. A person born into an upper-middle-class family will have greater advantages by the mere fact of birth than a person born into poverty.

One newspaper account criticized discussion by politicians on the subject of *equality* as "weasely", and thought that the term was politically correct and vague. Furthermore, when comparing *equality of opportunity* with *equality of outcome*, the sense was that the latter type was "worse" for society. Equality of outcome may be incorporated into a philosophy that ultimately seeks equality of opportunity. Moving towards a higher equality of outcome (albeit not perfectly equal) can lead to an environment more adept at providing equality of opportunity by eliminating conditions that restrict the possibility for members of society to fulfill their potential. For example, a child born in a poor, dangerous neighborhood with poor schools and little access to healthcare may be significantly disadvantaged in his attempts to maximize use of talents, no matter how fine his work ethic. Thus, even proponents of meritocracy may promote some level of equality of outcome in order to create a society capable of truly providing equality of opportunity.

While outcomes can usually be measured with a great degree of precision, it is much more difficult to measure the intangible nature of *opportunities*. That is one reason why many proponents of equal opportunity use measures of equality of outcome to judge success. Analyst Anne Phillips argued that the proper way to assess the effectiveness of the hard-to-measure concept of *equality of opportunity* is by the extent of the *equality of outcome*. Nevertheless, she described a single criterion of *equality of outcome* as problematic: the measure of "preference satisfaction" was "ideologically loaded" while other measures such as income or wealth were inadequate, and she advocated an approach which combined data about resources, occupations, and roles.

To the extent that inequalities can be passed from one generation to another through tangible gifts and wealth inheritance, some claim that equality of opportunity for children cannot be achieved without equality of outcome for parents. Moreover, access to social institutions is affected by equality of outcome and it's further claimed that rigging equality of outcome can be a way to prevent co-option of non-economic institutions important to social control and policy formation, such as the legal system, media or the electoral process, by powerful individuals or coalitions of wealthy people.

Purportedly, greater equality of outcome is likely to reduce relative poverty, leading to a more cohesive society. However, if taken to an extreme it may lead to greater absolute poverty if it negatively affects a country's GDP by damaging workers' sense of work ethic by destroying incentives to work harder. Critics of equality of outcome believe that it is more important to raise the standard of living of the poorest in absolute terms. Some critics additionally disagree with the concept of equality of outcome on philosophical grounds. Still others note that poor people of low social status often have a drive, hunger and ambition which ultimately lets them achieve better economic and social outcomes than their initially more advantaged rivals.

A related argument that is often encountered in education, especially in the debates on the grammar school in the United Kingdom and in the debates on gifted education in various countries, says that people by nature have differing levels of ability and initiative which result in some achieving better outcomes than others and it is therefore impossible to ensure equality of outcome without imposing inequality of opportunity.

The Concept in Political Argument

The concept of *equality of outcome* is an important one in battling between differing political positions, since the concept of *equality*, overall, was seen as positive and an important foundation which is "deeply embedded in the fabric of modern politics." There is much political jousting over what, exactly, equality means. It is not a new phenomenon; battling between so-called *haves* and *have-nots* has happened throughout human civilization, and was a focus of philosophers such as Aristotle in his treatise *Politics*. Analyst Julian Glover in *The Guardian* wrote that *equality* challenged both left-leaning and right-leaning positions, and suggested that the task of left-leaning advocates is to "understand the impossibility and undesirability of equality" while the task for right-leaning advocates was to "realise that a divided and hierarchical society cannot – in the best sense of that word – be fair."

- Conservatives. Conservatives believe in equality of opportunity not outcome. In their view, life is not fair, but that is how it is. They criticize attempts to try to fight poverty by redistributive methods as ineffective since more serious cultural and behavioral problems lock poor people into poverty. Sometimes right-leaning positions have been criticized by liberals for over-simplifying what is meant by the term *equality of outcome*, and for construing outcomes strictly to mean precisely equal amounts for everybody. Commentator Ed Rooksby in *The Guardian* criticized the right's tendency to oversimplify, and suggested that serious left-leaning advocates would not construe *equality* to mean "absolute equality of everything". Rooksby wrote that Marx favored the position described in the phrase "from each according to his ability, to each according to his need", and argued that this did not imply strict equality of things, but that it meant that people required "different things in different proportions in order to flourish."

- Libertarians and advocates of economic liberalism such as Friedrich Hayek and Milton Friedman tend to see equality of outcome negatively and argue that any effort to cause equal outcomes would necessarily and unfortunately involve coercion by government. Friedman wrote that striving for *equality of outcome* leaves most people "without equality and without opportunity."
- Liberals. Analyst Glenn Oliver suggested that liberals believed in "equality of opportunity and of outcome." One liberal position is that it is simplistic to define equality in strict outcomes since questions such as what is being equalized as well as huge differences in preferences and tastes and needs is considerable. They ask: exactly *what* is being equalized? In the 1960s in the United States, mainstream liberal president Lyndon Johnson, examining the plight of African Americans locked in poverty, argued for ending policies which promoted segregation and discrimination as well as steps to end "economic injustice" by turning "equality of opportunity into equality of outcome," that is, with programs to transfer wealth in varying amounts. Fairness is emphasized; one writer expounding a centrist position wrote "people would neither be left to fend for themselves nor guaranteed equality of outcome – they would be given the tools they needed to achieve the American dream if they worked hard." There has been cynicism expressed in the media that neither side, including mainstream political positions, wants to do anything substantive, but that the nebulous term *fairness* is used to cloak the inactivity because it is difficult to measure what, in fact, "fairness" means. Julian Glover wrote that fairness "compels no action" and compared it to an "atmospheric ideal, an invisible gas, a miasma," and to use an expression by Churchill, a "happy thought."
- Social democrats champion greater equality of outcome and opportunities within capitalism, usually promoted through redistributive social policies like progressive taxation and the provision of universal public services.
- Socialists often believe in both "inequality of opportunity and equality of outcome" according to Oliver. They often see greater *equality of outcome* as a positive long-term goal to be achieved, so that individuals have equal access to the means of production and consumption. Only a small minority of socialist theories advocate complete economic equality of outcome. The vast majority of socialists view an ideal economy as one where remuneration is at least somewhat proportional to the degree of effort and personal sacrifice expended by individuals in the productive process. This latter concept was expressed by Karl Marx's famous maxim: *To each according to his contribution.*

References

- Ceadel, Martin (1996). The Origins of War Prevention: The British Peace Movement and International Relations, 1730–1854. Oxford University Press,. Retrieved 2013-02-07
- The Abolition of War: the Peace Movement in Britain, 1914–1919 by Keith Robbins. University of Wales Press, 1976. ISBN 978-0-7083-0622-2 (p.10)
- "War and the Iliad". The New York Review of books. Archived from the original on May 1, 2008. Retrieved 29 September 2009
- Lest Innocent Blood Be Shed: The Story of Le Chambon and How Goodness Happened There Philip P. Hallie, (1979) New York: Harper & Row, ISBN 0-06-011701-X
- Ceadel, Martin (1996). The Origins of War Prevention: The British Peace Movement and International Relations, 1730–1854. Oxford University Press,. Retrieved 2013-02-07
- Phillips, A. (2004). Defending equality of outcome, Journal of Political Philosophy, 12/1, 2004, pp. 1–19, Defending Equality of Outcome, Accessed July 12, 2013
- Linus Pauling (October 10, 1963). "Notes by Linus Pauling. October 10, 1963.". Oregon State University Libraries Special Collections. Retrieved December 13, 2007
- In Solitary Witness: The Life and Death of Franz Jägerstätter by Gordon Zahn. Springfield, Illinois: Templegate Publishers. ISBN 0-87243-141-X
- "For a Middle East free of all Weapons of Mass Destruction". Campaign Against Sanctions and Military Intervention in Iran. 2007-08-06. Retrieved 2007-11-03
- Anne Phillips (2004). "Defending Equality of Outcome". Journal of Political Philosophy. pp. 1–19. Retrieved 2011-07-15
- Mark E. Rushefsky (2008). "Public Policy in the United States: At the Dawn of the Twenty-First Century". M. E. Sharpe Inc. Retrieved 2011-07-15

5

Green Politics and the Environment

Ecotheology is a theory that studies the relation between religion and nature. Along with the degradation of nature it also focuses on ecosystem management. The topics discussed in the chapter are of great importance to broaden the existing knowledge on green politics.

Ecotheology

Ecotheology is a form of constructive theology that focuses on the interrelationships of religion and nature, particularly in the light of environmental concerns. Ecotheology generally starts from the premise that a relationship exists between human religious/spiritual worldviews and the degradation of nature. It explores the interaction between ecological values, such as sustainability, and the human domination of nature. The movement has produced numerous religious-environmental projects around the world.

The burgeoning awareness of environmental crisis has led to widespread religious reflection on the human relationship with the earth. Such reflection has strong precedents in most religious traditions in the realms of ethics and cosmology, and can be seen as a subset or corollary to the theology of nature.

It is important to keep in mind that ecotheology explores not only the relationship between religion and nature in terms of degradation of nature, but also in terms of ecosystem management in general. Specifically, ecotheology seeks not only to identify prominent issues within the relationship between nature and religion, but also to outline potential solutions. This is of particular importance because many supporters and contributors of ecotheology argue that science and education are simply not enough to inspire the change necessary in our current environmental crisis.

Background

The relationship of theology to the modern ecological crisis became an intense issue of debate in Western academia in 1967, following the publication of the article, "The Historical Roots of Our Ecological Crisis, " by Lynn White, Jr., Professor of History at the University of California at Los Angeles. In this work, White puts forward a theory that the Christian model of human dominion over nature has led to environmental devastation, providing a voice for "The Ecological Complaint".

In 1973, theologian Jack Rogers published an article in which he surveyed the published studies of approximately twelve theologians which had appeared since White's article. They reflect the search for "an appropriate theological model" which adequately assesses the biblical data regarding the relationship between God, humans, and nature.

Precedents in Religious Thought

Some scholars argue that Christians actually helped bring about the current global environmental crisis by instructing followers that God, and by extension mankind, transcends nature. Much of the development of ecotheology as a theological discourse was in response to this argument, which has been called "The Ecological Complaint". Defendants of this perspective essentially claim that Christianity promotes the idea of human dominion over nature, treating nature itself as a tool to be used and even exploited for survival and prosperity.

However, Christianity has often been viewed as the source of positive values towards the environment, and there are many voices within the Christian tradition whose vision embraces the well-being of the earth and all creatures. While Francis of Assisi is one of the more obvious influences on Christian ecotheology, there are many theologians and teachers, such as Isaac of Nineveh and Seraphim of Sarov, whose work has profound implications for Christian thinkers. Many of these are less well known in the West because their primary influence has been on the Orthodox Church rather than the Roman Catholic Church.

The significance of indigenous traditions for the development of ecotheology can also not be understated. Systems of Traditional Ecological Knowledge, in combination with modern scientific methods of ecosystem management, are steadily gaining interest as environmental activists realize the importance of locally invested groups.

Further Exploration

Christian ecotheology draws on the writings of such authors as Jesuit priest and paleontologist Pierre Teilhard de Chardin, philosopher Alfred North Whitehead, and Passionist priest and historian Thomas Berry. It is well represented in Protestantism by John B. Cobb, Jr., Jürgen Moltmann, and Michael Dowd; in ecofeminism by feminist theologians Rosemary Radford Ruether, Catherine Keller, and Sallie McFague; in Roman Catholicism by John F. Haught; and in Orthodoxy by Elizabeth Theokritoff and George Nalunnakkal (currently Bishop Geevarghese Mor Coorilose of the Jacobite Syrian Christian Church). Besides works on theology per se, interpreters of the ecological significance of scripture, such as Ellen Davis, also play an important role.

Creation Spirituality is another important expression of ecotheology that has been developed and popularized by Matthew Fox, a former Catholic Dominican friar turned Episcopal priest.

Abraham Joshua Heschel and Martin Buber, both Jewish philosophers, have also left their mark on Christian ecotheology, and provide significant inspiration for Jewish ecotheology. The most recent and most complete expression of Jewish ecotheology to date can be found in David Mevorach Seidenberg's work on Kabbalah and ecology.

Hindu ecotheology includes writers such as Vandana Shiva. Seyyid Hossein Nasr, a Perennialist scholar and Persian Sufi philosopher, was one of the earlier Muslim voices calling for a re-evaluation of the Western relationship to nature.

Elisabet Sahtouris is an evolutionary biologist and futurist who promotes a vision she believes will result in the sustainable health and well-being of humanity within the larger living systems of Earth and the cosmos. She is a lecturer in Gaia Theory and a co-worker with James Lovelock and Lynn Margulis.

Annie Dillard, Pulitzer Prize-winning American author, also combined observations on nature and philosophical explorations in several ecotheological writings, including *Pilgrim at Tinker Creek*.

Valerie Brown is a science and environmental journalist based in Portland, Oregon, whose work has appeared in *Environmental Health Perspectives, 21stC*, and other publications. She writes regularly about ecotheology.

Terry Tempest Williams is a Mormon writer and conservationist who sensitively and imaginatively explores ecotheology in her very personal writing.

The majority of the content of *Indians of the Americas*, by former Bureau of Indian Affairs head John Collier, concerns the link between ecological sustainability and religion among Native North and South Americans.

An important book on perhaps the first ecotheologian, Paul Tillich—who was writing on this issue long before the term ecotheology was even coined—is "The Ecotheology of Paul Tillich: Spiritual Roots of Environmental Ethics" (Santa Barbara: Barred Owl Books, 2009).

Environmental Technology

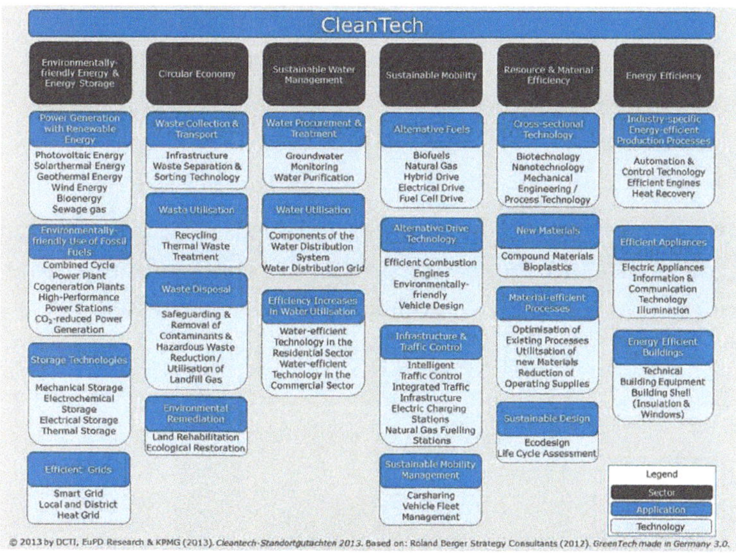

CleanTech

Environmental technology (*envirotech*), green technology (*greentech*) or clean technology (*cleantech*) is the application of one or more of environmental science, green chemistry, environmental monitoring and electronic devices to monitor, model and conserve the natural environment and resources, and to curb the negative impacts of human involvement. The term is also used to describe sustainable energy generation technologies such as photovoltaics, wind turbines, bioreactors, etc. Sustainable development is the core of *environmental technologies*. The term *environmental technologies* is also used to describe a class of electronic devices that can promote sustainable management of resources.

Sustainable urban design and innovation: Photovoltaic ombrière SUDI is an autonomous and mobile station that replenishes energy for electric vehicles using solar energy.

Examples

- Biofiltration
- Bioreactor
- Bioremediation
- Desalination
- Doubly fed electric machine
- Energy conservation
- Energy saving modules
- Electric vehicles
- Wave energy
- Green computing
- Hydroelectricity
- Wind power
- Wind turbine
- Hydrogen fuel cell
- Ocean thermal energy conversion
- Solar power
- Photovoltaic
- Thermal depolymerization
- Composting toilet
- Pyrolysis

Renewable Energy

Renewable energy is the energy that can be replenished easily. For years we have been using sources such as wood, sun, water, etc. for means for producing energy. Energy that can be produced by natural objects like wood, sun, wind, etc. is considered to be renewable.

A view across a reverse osmosis desalination plant in Nepal.

Water Purification

Water purification: The whole idea/concept of having dirt/germ/pollution free water flowing throughout the environment. Many other phenomena lead from this concept of purification of water. Water pollution is the main enemy of this concept, and various campaigns and activists have been organized around the world to help purify water.

Air Purification

Air purification: Basic and common green plants can be grown indoors to keep air fresh because all plants remove CO_2 and convert it into oxygen. The best examples are: *Dypsis lutescens*, *Sansevieria trifasciata*, and *Epipremnum aureum*.

Sewage Treatment

Sewage treatment is conceptually similar to water purification. Sewage treatments are very important as they purify water per levels of its pollution. The most polluted water is not used for anything, and the least polluted water is supplied to places where water is used affluently. It may lead to various other concepts of environmental protection, sustainability etc.

Environmental Remediation

Environmental remediation is the removal of pollutants or contaminants for the general protection of the environment. This is accomplished by various chemical, biological, and bulk movements. (encyclopedia of medical concepts)

Solid waste management

Net Zero Court zero emissions office building prototype in St. Louis, Missouri

Solid waste management is the purification, consumption, reuse, disposal and treatment of solid waste that is undertaken by the government or the ruling bodies of a city/town.

Egain Forecasting

Egain forecasting is a method using forecasting technology to predict the future weather's impact on a building. By adjusting the heat based on the weather forecast, the system eliminates redundant use of heat, thus reducing the energy consumption and the emission of greenhouse gases.

Energy Conservation

Energy conservation is the utilization of devices that require smaller amounts of energy in order to reduce the consumption of electricity. Reducing the use of electricity causes less fossil fuels to be burned to provide that electricity.

Alternative and Clean Power

The Tesla Roadster was the first all-electric sports car for sale and in serial production. It can be plugged into conventional outlets and can be charged fully or partially on renewable energy, including solar, hydroelectric, geothermal or wind power.

Principles:

- Green syndicalism
- Sustainability
- Sustainable design
- Sustainable engineering

Scientists continue to search for clean energy alternatives to our current power production methods. Some technologies such as anaerobic digestion produce renewable energy from waste materials. The global reduction of greenhouse gases is dependent on the adoption of energy conservation technologies at industrial level as well as this clean energy generation. That includes using unleaded gasoline, solar energy and alternative fuel vehicles, including plug-in hybrid and hybrid electric vehicles.

Since industrial use of energy accounts for 51% of worldwide energy usage improving energy efficiency in this field is a top priority for environmental technology companies around the globe. Advanced energy efficient electric motor (and electric generator) technology that are cost effective to encourage their application, such as the brushless wound-rotor doubly fed electric machine and energy saving module, can reduce the amount of carbon dioxide (CO_2) and sulfur dioxide (SO_2) that would otherwise be introduced to the atmosphere, if electricity is generated using fossil fuels. Greasestock is an event held yearly in Yorktown Heights, New York which is one of the largest showcases of environmental technology in the United States.

Education

Courses aimed at developing graduates with specific skills in environmental systems or environmental technology are becoming more common and fall into three broads classes:

- *Environmental Engineering or Environmental Systems* courses oriented towards a civil engineering approach in which structures and the landscape are constructed to blend with or protect the environment;
- *Environmental chemistry, sustainable chemistry* or *environmental chemical engineering* courses oriented towards understanding the effects (good and bad) of chemicals in the environment. Such awards can focus on mining processes, pollutants and commonly also cover biochemical processes;
- *Environmental technology* courses oriented towards producing electronic, electrical or electrotechnology graduates capable of developing devices and artefacts able to monitor, measure, model and control environmental impact, including monitoring and managing energy generation from renewable sources, and developing novel energy generation technologies.

Political Ecology

Political ecology studies the complex interaction between economics, politics, technology, social tradition and the biological environment. These terraced rice fields in Yunnan, China, evidence how the environment is shaped by and shapes economy and society.

Political ecology is the study of the relationships between political, economic and social factors with environmental issues and changes. Political ecology differs from apolitical ecological studies by politicizing environmental issues and phenomena.

The academic discipline offers wide-ranging studies integrating ecological social sciences with political economy in topics such as degradation and marginalization, environmental conflict, conservation and control, and environmental identities and social movements.

Origins

The term "political ecology" was first coined by Frank Thone in an article published in 1935. It has been widely used since then in the context of human geography and human ecology, but with no systematic definition. Anthropologist Eric R. Wolf gave it a second life in 1972 in an article entitled "Ownership and Political Ecology," in which he discusses how local rules of ownership and inheritance "mediate between the pressures emanating from the larger society and the exigencies of the local ecosystem", but did not develop the concept further Other origins include other early works of Eric R. Wolf, Hans Magnus Enzensberger and others in the 1970s and 1980s.

The origins of the field in the 1970s and 1980s were a result of the development of development geography and cultural ecology., particularly the work of Piers Blaikie on the sociopolitical origins of soil erosion. Historically, political ecology has focused on phenomena in and affecting the developing world; since the field's inception, "research has sought primarily to understand the political dynamics surrounding material and discursive struggles over the environment in the third world".

Scholars in political ecology are drawn from a variety of academic disciplines, including geography, anthropology, development studies, political science, sociology, forestry, and environmental history.

Overview

Political ecology's broad scope and interdisciplinary nature lends itself to multiple definitions and understandings. However, common assumptions across the field give the term relevance. Raymond L. Bryant and Sinéad Bailey developed three fundamental assumptions in practicing political ecology:

- First, costs and benefits associated with environmental change are distributed unequally. Changes in the environment do not affect society in a homogenous way: political, social, and economic differences account for uneven distribution of costs and benefits.
- Second, this unequal distribution inevitably reinforces or reduces existing social and economic inequalities - "any change in environmental conditions must affect the political and economic status quo."
- Third, the unequal distribution of costs and benefits and the reinforcing or reducing of pre-existing inequalities has political implications in terms of the altered power relationships that then result.

In addition, political ecology attempts to provide critiques and alternatives in the interplay of the environment and political, economic and social factors. Paul Robbins asserts that the discipline has a "normative understanding that there are very likely better, less coercive, less exploitative, and more sustainable ways of doing things".

From these assumptions, political ecology can be used to:

- inform policymakers and organizations of the complexities surrounding environment and development, thereby contributing to better environmental governance.

- understand the decisions that communities make about the natural environment in the context of their political environment, economic pressure, and societal regulations
- look at how unequal relations in and among societies affect the natural environment, especially in context of government policy.

Scope and Influences

Political ecology's movement as a field since its inception in the 1970s has complicated its scope and goals. Through the discipline's history, certain influences have grown more and less influential in determining the focus of study. Peter A. Walker traces the importance of the ecological sciences in political ecology. He points to the transition, for many critics, from a 'structuralist' approach through the 1970s and 1980s, in which ecology maintains a key position in the discipline, to a 'poststructuralist' approach with an emphasis on the 'politics' in political ecology. This turn has raised questions as to the differentiation with environmental politics as well as the field's use of the term of 'ecology'. Political ecological research has shifted from investigating political influence on the earth's surface to the focus on spatial-ecological influences on politics and power—a scope reminiscent of environmental politics.

Much has been drawn from cultural ecology, a form of analysis that showed how culture depends upon, and is influenced by, the material conditions of society (political ecology has largely eclipsed cultural ecology as a form of analysis according to Walker.) As Walker states, "whereas cultural ecology and systems theory emphasize[s] adaptation and homeostasis, political ecology emphasize[s] the role of political economy as a force of maladaptation and instability".

Political ecologists often use political economy frameworks to analyze environmental issues. Early and prominent examples of this were *The Political Economy of Soil Erosion in Developing Countries* by Piers Blaikie in 1985, which traced land degradation in Africa to colonial policies of land appropriation, rather than over-exploitation by African farmers, and *Silent Violence: Food, Famine and Peasantry in Northern Nigeria* by Michael Watts in 1983, which traced the famine in northern Nigeria during the 1970s to the effects of colonialism, rather than an inevitable consequence of the drought in the Sahel.

Relationship to Anthropology and Geography

Originating in the 18th and 19th centuries with philosophers such as Adam Smith, Karl Marx, and Thomas Malthus, political economy attempted to explain the relationships between economic production and political processes. It tended toward overly structuralist explanations, focusing on the role of individual economic relationships in the maintenance of social order. Eric Wolf used political economy in a neo-Marxist framework which began addressing the role of local cultures as a part of the world capitalist system, refusing to see those cultures as "primitive isolates". But environmental effects on political and economic processes were under-emphasised.

Conversely, Julian Steward and Roy Rappaport's theories of cultural ecology are sometimes credited with shifting the functionalist-oriented anthropology of the 1950s and 1960s and incorporating ecology and environment into ethnographic study.

Geographers and anthropologists worked with the strengths of both to form the basis of political ecology. PE focuses on issues of power, recognizing the importance of explaining environmental impacts on cultural processes without separating out political and economic contexts. The actual term "political ecology" was first used in 1988 in a geographical publication by Tom Bassett.

The application of political ecology in the work of anthropologists and geographers differs. While any approach will take both the political/economic and the ecological into account,the emphasis can be unequal. Some, such as geographer Michael Watts, focus on how the assertion of power impacts on access to environmental resources. His approach tends to see environmental harm as both a cause and an effect of "social marginalization".

Political ecology has strengths and weaknesses. At its core, it contextualizes political and ecological explanations of human behavior. But as Walker points out, it has failed to offer "compelling counter-narratives" to "widely influential and popular yet deeply flawed and unapologetic neo-Malthusian rants such as Robert Kaplan's (1994) 'The coming anarchy' and Jared Diamond's (2005) *Collapse* (385). Ultimately, applying political ecology to policy decisions – especially in the US and Western Europe – will remain problematic as long as there is a resistance to Marxist and neo-Marxist theory.

Andrew Vayda and Bradley Walters (1999) criticize political ecologists for pre-supposing "the importance ... of certain kinds of political factors in the explanation of environmental changes" (167). Vayda and Walter's response to overly political approaches in political ecology is to encourage what they call "event ecology", focusing on human responses to environmental events without presupposing the impact of political processes on environmental events. The critique has not been taken up widely.

Relationship to Conservation

There is a divergence of ideas between conservation science and political ecology. With conservationists establishing protected areas to conserve biodiversity, "political ecologists have devoted some energy to the study of protected areas, which is unsurprising given political ecology's overall interest in forms of access to, and control over resources". The arguments against enclosure of land for conservation is that it harms local people and their livelihood systems, by denying them access. As Dove and Carpenter state, "indigenous people have important environmental knowledge which could contribute to conservation". The objection by political ecologists is that land use regulations are made by NGOs and the government, denying access, denying the ability of local people to conserve species and areas themselves, and rendering them more vulnerable through dispossession.

In a few cases, perhaps especially tragic local groups have been displaced to create national parks and reserves to 'conserve' the forest. Fortunately, most conservation bodies are now aware that, if a group has been using and managing a forest for several thousand years, throwing it off the land is more apt to destroy the forest ecosystem than to preserve it. (Sutton 2004: 302)

Green Development

Green development is a real estate development concept that carefully considers social and environmental impacts of development. It is defined by three sub-categories: environmental responsiveness, resource efficiency, and community and cultural sensitivity. Environmental responsiveness respects the intrinsic value of nature, and minimizes damage to an ecosystem. Environmental efficiency refers to the use of fewer resources to conserve energy and the environment. Community and cultural sensitivity recognizes the unique cultural values that each community hosts and carefully considers them in real estate development, unlike more discernable signs of sustainability, like solar energy, (solar panels are more visibly "green" than the use of local materials). Green development manifests itself in various forms, however it is generally based on solution multipliers: features of a project that provide additional benefits, which ultimately reduce the projects' environmental impacts.

The above green development concepts have been derived from the all-inclusive tome released by The Rocky Mountain Institute, entitled Green Development: Integrating Ecology and Real Estate.

History

Green development emerged as a result of the environmental movement in the 1970s. In real estate however, the term commenced in 1897 with a report from the World Commission on Environment and Development, entitled Our Common Future. The report includes 16 principles of environmental management, designed to foster green development. It also discusses the traditional model of macroeconomic growth, and its disregard for environmental consequences. From this initial movement, the future experienced a back-and-forth relationship with green methodologies; environmental issues often came second behind purely economic factors. Incessant environmental concern and legislation, including in real estate development began to emerge, i.e. Green development. However, a common concern of green development is that it may increase costs and time, hence the ongoing argument of whether green strategies can be sustainable as well as economically stimulating. National environmental attention has since, worked its way down to real estate developers, and become an increasing priority. Developers today, must work within the parameters of legislation that now considers the environmental implications of development.

Legislation

In response to increasing public concern regarding environmental issues, the government adopted a series of environmental controls in the real estate market. For example, in California, such legislation included the National Environmental Policy Act, the NEPA process, the Clean Air Act, the Clean Water Act, the Coastal Zone Management Act, and the Comprehensive Environmental Response, Compensation, and Liability Act (CERCLA). The National Environmental Policy Act, signed in 1970, changed how federal agencies made decisions because it required them to propose environmental analysis before starting a project. The Clean Air Act (1970) requires the EPA to set national models for clean air. The Clean Water Act (1972) was designed to minimize pollution in natural bodies of water, and also to create bodies of water that are clean enough for recreational activities such as fishing or swimming. The Coastal Zone Management Act (1972) manages the nations coastal resources such as Great Lakes and estuaries. The Comprehensive Environmental Response, Compensation, and Liability Act (CERCLA), is also known as "Superfund." It is commonly referred to as "Superfund" because it comprises two trust funds that provide help to improve areas that have been polluted by hazardous waste. The Superfund Amendments and Reauthorization Act of 1986 allows the government to lien a property that is being cleaned up. Additionally, the

California Environmental Quality Act (CEQA) is California's most comprehensive piece of legislation regarding the environment. This act applies to all decisions made by cities and counties in California, and includes the mandate of an Environmental Impact Report (EIR), to both public and private projects. Subsequently, any new real estate development is subject to a detailed environmental analysis before starting a project.

California's Senate Bill (SB) 375 (2008) is another piece of legislation that promotes green development. It aims to achieve California's climate goals via more efficient land use and development patterns. More specifically, SB 375 seeks to reduce greenhouse gas (GHG) emissions through close coordination between land-use and transportation. One way this is achieved is through demand-side measures. This strategy would decrease driving demand, and therefore reduce vehicle miles traveled (VMT), and ultimately reduce GHG emissions. For example, a demand-side policy component may include placing public transit stops near development, in order to maximize walkability.

Additionally, California state law requires minimum energy conservation levels for all new and/or existing development projects. The seller of a home is required to include information regarding energy conservation retrofitting and thermal insulation in the sales contract.

Green Development in Practice

Broad Development Patterns

Senate Bill 375 demonstrates an urban planning strategy called growth management. It is defined by a close coordination between land-use controls and capital investment and heavily motivated by environmental issues. It is defined by "the regulation of the amount, timing, location, and character of development." As the name may suggest, growth management may not imply limiting any growth. "Growth control" carries the connotation of managing or limiting growth, and "no growth" would indicate stopping growth all together. Moreover, Growth Management requires the cooperation of all three of these connotations.

Urban Growth Boundaries (UBG's), are popular growth management strategies. They are designed to encourage growth within a given boundary and discourage it outside the boundary. The goal of the UGB is to promote dense development, in order to decrease urban sprawl. This growth management technique ultimately seeks to revitalize central cities, and create vibrant, walk-able spaces for community development.

These clustered development patterns are solution multipliers. Reducing demand for infrastructure can save money and resources. These multipliers can increase walkability, which fosters social interaction and community togetherness.

Intelligent Building

In the US, commercial and residential buildings are the highest consumers of electricity and HVAC systems comprise a large portion of this usage. In fact, the US Department of Energy projects that 70% of the electricity used in the US is from buildings. Intelligent Building methods such as Occupancy Detection Systems, Wireless Sensor Networks, and HVAC Control Systems aim to more efficient energy usage. A team of researchers at the University of San Diego, project that their smart building automation systems will save 10%-15% in building energy.

Case Studies

The Holly Street Village Apartments

The city of Pasadena has recently adopted a general plan based on seven guiding principles: community needs and quality of life, preservation of Pasadena's historic character, economic vitality, a healthy family community, lack of need for automobiles, promoted as a cultural, scientific, corporate, entertainment and educational center for the region, and community participation.

This project included environmental concern and social considerations in the process of construction. The Holly Street Village Apartments in Pasadena address several of the principles outlined in Pasadena's general plan. It incorporates mixed-use development with ground-floor retail center including a deli, a convenience store and an art gallery. Also, the Holly Street Village Apartments is located near a light rail station. The goal of these strategies is to reduce the demand for automobiles, and making it easier for people to use public transportation.

Inn of the Anazasi

Zimmer Associates International, a real estate development firm, completed the Inn of the Anasazi in Santa Fe, New Mexico in 1991. Robert Zimmer (co-founder) and his partners, Steve Conger and Michael Fuller, set a goal to construct a building that would, "showcase energy- and resource-saving technologies, strengthen local community, offer first class elegance, and financially reward its participants." The interior design of the hotel pays respect to the ancient Anasazi Indians, including locally crafted furniture, hand-made rugs, and Native American, Hispanic and cowboy wall art. The use of adobe on the exterior of the hotel includes the historic pueblo style. Also, Zimmer and his partners repurposed a steel-framed building that had previously been used in the 1960s as a juvenile detention center, instead of starting the project from the beginning. Other "green" characteristics of this Santa Fe hotel are skylights, energy-efficient lighting, and water-saving fixtures. Also, the Inn of the Anasazi stimulates the regional economy by purchasing locally grown organic food from Hispanic farmers. Lastly the Inn encourage that the staff participate in local nonprofit organizations and events that sponsor diverse local cultures. The Inn of the Anasazi integrates "social and environmental goals with financial considerations..."

Taipei 101

Taipei 101 is a 1,667-foot tall skyscraper located in Taiwan, and has received LEED (Leadership in Energy and Environmental Design) certification from the US Green Building Council as the highest score in history. Taipei is an acronym for "technology," "art," "innovation," "people," "environment," and "identity. Dr. Hubert Keiber, CEO of Siemens Building Automation in Taipei, stresses energy efficiency on the grounds that energy is the single largest expense for commercial buildings. Since 2011, the Taipei 101 has become the world's most environmentally responsible skyscraper, reducing water use, energy use, and carbon emissions all by 10%.

Boulder, Colorado

Growth management/limitation (discussed previously) manifests itself in Boulder, Colorado. The city of Boulder very tightly restricts housing development by limiting housing permits to 400 per year, which is 1 percent of the city's total housing stock. Additionally, the city has purchased land outside of the city limits, designated for permanent, green open space. This 400-unit cap seriously hinders population growth in the city. This shortage of housing has several repercussions. First of all, it increases housing prices. Also, because Boulder restricts housing development more than it does commercial development, the number of available workers in Boulder grows faster than the housing stock. This results in many workers who commute from beyond the city limits.

Controversy

A common critique of green development is that it negatively affects the way real estate developers do business, as it can increase cost and create delay. For example, becoming LEED certified (Leaders in Energy and Environmental Design), can contribute to additional costs. This includes additional building design and construction fees, interior design and construction fees, building operations and maintenance fees, neighborhood development fees, home and campus fees, and volume program fees. Additionally, green development has been critiqued on a residential level. High-performance homes have proven to save energy in the long run, but they rapidly increase up-front capital costs, via tankless water heaters, radiant barriers and reflective insulation systems, and high efficiency air conditioning systems. Also, developers are often unable to develop on certain portions of land due to conservation easements. These easements are purchased by governments or non-governmental organizations, in order to "preserve land in its natural, scenic, agricultural, historical, forested, or open-space condition." However, some developers argue that conservation easements leave landowners with limited uses.

Green Building

US EPA Kansas City Science & Technology Center. This facility features the following green attributes:

- LEED 2.0 Gold certified
- Green Power
- Native Landscaping

Green building (also known as green construction or sustainable building) refers to both a structure and the application of processes that are environmentally responsible and resource-efficient throughout a building's life-cycle: from planning to design, construction, operation, maintenance, renovation, and demolition. This requires close cooperation of the contractor, the architects, the engineers, and the client at all project stages. The Green Building practice expands and complements the classical building design concerns of economy, utility, durability, and comfort.

Leadership in Energy and Environmental Design (LEED) is a set of rating systems for the design, construction, operation, and maintenance of green buildings which was Developed by the U.S. Green Building Council. Other certificates system that confirms the sustainability of buildings is the British BREEAM (Building Research Establishment Environmental Assessment Method) for buildings and large scale developments. Currently, World Green Building Council is conducting research on the effects of green buildings on the health and productivity of their users and is working with World Bank to promote Green Buildings in Emerging Markets through EDGE Excellence in Design for Greater Efficiencies Market Transformation Program and certification.

Although new technologies are constantly being developed to complement current practices in creating greener structures, the common objective of green buildings is to reduce the overall impact of the built environment on human health and the natural environment by:

- Efficiently using energy, water, and other resources
- Protecting occupant health and improving employee productivity
- Reducing waste, pollution and environmental degradation

A similar concept is natural building, which is usually on a smaller scale and tends to focus on the use of natural materials that are available locally. Other related topics include sustainable design and green architecture. Sustainability may be defined as meeting the needs of present generations without compromising the ability of future generations to meet their needs. Although some green building programs don't address the issue of the retrofitting existing homes, others do, especially through public schemes for energy efficient refurbishment. Green construction principles can easily be applied to retrofit work as well as new construction.

A 2009 report by the U.S. General Services Administration found 12 sustainably-designed buildings that cost less to operate and have excellent energy performance. In addition, occupants were overall more satisfied with the building than those in typical commercial buildings.These are eco-friendly buildings.

Reducing Environmental Impact

Globally buildings are responsible for a huge share of energy, electricity, water and materials consumption. The building sector has the greatest potential to deliver significant cuts in emissions at little or no cost. Buildings account for 18% of global emissions today, or the equivalent of 9 billion tonnes of CO_2 annually. If new technologies in construction are not adopted during this time of rapid growth, emissions could double by 2050, according to the United Nations Environment Program.Green building practices aim to reduce the environmental impact of building. Since construction almost always degrades a building site, not building at all is preferable to green building, in terms of reducing environmental impact. The second rule is that every building should be as small as possible. The third rule is not to contribute to sprawl, even if the most energy-efficient, environmentally sound methods are used in design and construction.

Hanging gardens of One Central Park, Sydney

Buildings account for a large amount of land. According to the National Resources Inventory, approximately 107 million acres (430,000 km[2]) of land in the United States are developed. The International Energy Agency released a publication that estimated that existing buildings are responsible for more than 40% of the world's total primary energy consumption and for 24% of global carbon dioxide emissions.

Goals Of Green Building

Blu Homes mkSolaire, a green building designed by Michelle Kaufmann.

The concept of sustainable development can be traced to the energy (especially fossil oil) crisis and environmental pollution concerns of the 1960s and 1970s. The Rachel Carson book, "Silent Spring", published in 1962, is considered to be one of the first initial efforts to describe sustainable development as related to green building. The green building movement in the U.S. originated from the need and desire for more energy efficient and environmentally friendly construction practices. There are a number of motives for building green, including environmental, economic, and social benefits. However, modern sustainability initiatives call for an integrated and synergistic design to both new construction and in the retrofitting of existing structures. Also known as sustainable design, this approach integrates the building life-cycle with each green practice employed with a design-purpose to create a synergy among the practices used.

Green building brings together a vast array of practices, techniques, and skills to reduce and ultimately eliminate the impacts of buildings on the environment and human health. It often emphasizes taking advantage of renewable resources, e.g., using sunlight through passive solar, active solar, and photovoltaic equipment, and using plants and trees through green roofs, rain gardens, and reduction of rainwater run-off. Many other techniques are used, such as using low-impact building materials or using packed gravel or permeable concrete instead of conventional concrete or asphalt to enhance replenishment of ground water.

Taipei 101, the tallest and largest green building of LEED Platinum certification in the world since 2011.

While the practices or technologies employed in green building are constantly evolving and may differ from region to region, fundamental principles persist from which the method is derived: siting and structure design efficiency, energy efficiency, water efficiency, materials efficiency, indoor environmental quality enhancement, operations and maintenance optimization and waste and toxics reduction. The essence of green building is an optimization of one or more of these principles. Also, with the proper synergistic design, individual green building technologies may work together to produce a greater cumulative effect.

On the aesthetic side of green architecture or sustainable design is the philosophy of designing a building that is in harmony with the natural features and resources surrounding the site. There are several key steps in designing sustainable buildings: specify 'green' building materials from local sources, reduce loads, optimize systems, and generate on-site renewable energy.

Life Cycle Assessment

A life cycle assessment (LCA) can help avoid a narrow outlook on environmental, social and economic concerns by assessing a full range of impacts associated with all cradle-to-grave stages of a process: from extraction of raw materials through materials processing, manufacture, distribution, use, repair and maintenance, and disposal or recycling. Impacts taken into account include (among others) embodied energy, global warming potential, resource use, air pollution, water pollution, and waste.

In terms of green building, the last few years have seen a shift away from a *prescriptive* approach, which assumes that certain prescribed practices are better for the environment, toward the scientific evaluation of actual performance through LCA.

Although LCA is widely recognized as the best way to evaluate the environmental impacts of buildings (ISO 14040 provides a recognized LCA methodology), it is not yet a consistent requirement of green building rating systems and codes, despite the fact that embodied energy and other life cycle impacts are critical to the design of environmentally responsible buildings.

In North America, LCA is rewarded to some extent in the Green Globes® rating system, and is part of the new American National Standard based on Green Globes, *ANSI/GBI 01-2010: Green Building Protocol for Commercial Buildings*. LCA is also included as a pilot credit in the LEED system, though a decision has not been made as to whether it will be incorporated fully into the next major revision. The state of California also included LCA as a voluntary measure in its 2010 draft *Green Building Standards Code.*

Although LCA is often perceived as overly complex and time consuming for regular use by design professionals, research organizations such as BRE in the UK and the Athena Sustainable Materials Institute in North America are working to make it more accessible.

In the UK, the BRE *Green Guide to Specifications* offers ratings for 1,500 building materials based on LCA.

In North America, the ATHENA® *EcoCalculator for Assemblies* provides LCA results for several hundred common building assembles based on data generated by its more complex parent software, the ATHENA® *Impact Estimator for Buildings.* Athena software tools are especially useful early in the design process when material choices have far-reaching implications for overall environmental impact. They allow designers to experiment with different material mixes to achieve the most effective combination.

Siting and Structure Design Efficiency

The foundation of any construction project is rooted in the concept and design stages. The concept stage, in fact, is one of the major steps in a project life cycle, as it has the largest impact on cost and performance. In designing environmentally optimal buildings, the objective is to minimize the total environmental impact associated with all life-cycle stages of the building project.

Exterior Light Shelves - Green Office Building, Denver Colorado

However, building as a process is not as streamlined as an industrial process, and varies from one building to the other, never repeating itself identically. In addition, buildings are much more complex products, composed of a multitude of materials and components each constituting various design variables to be decided at the design stage. A variation of every design variable may affect the environment during all the building's relevant life-cycle stages.

Energy Efficiency

An eco-house at Findhorn Ecovillage with a turf roof and solar panels

Green buildings often include measures to reduce energy consumption – both the embodied energy required to extract, process, transport and install building materials and operating energy to provide services such as heating and power for equipment.

As high-performance buildings use less operating energy, embodied energy has assumed much greater importance – and may make up as much as 30% of the overall life cycle energy consumption. Studies such as the U.S. LCI Database Project show buildings built primarily with wood will have a lower embodied energy than those built primarily with brick, concrete, or steel.

To reduce operating energy use, designers use details that reduce air leakage through the building envelope (the barrier between conditioned and unconditioned space). They also specify high-performance windows and extra insulation in walls, ceilings, and floors. Another strategy, passive solar building design, is often implemented in low-energy homes. Designers orient windows and walls and place awnings, porches, and trees to shade windows and roofs during the summer while maximizing solar gain in the winter. In addition, effective window placement (daylighting) can provide more natural light and lessen the need for electric lighting during the day. Solar water heating further reduces energy costs.

Onsite generation of renewable energy through solar power, wind power, hydro power, or biomass can significantly reduce the environmental impact of the building. Power generation is generally the most expensive feature to add to a building.

Water Efficiency

Reducing water consumption and protecting water quality are key objectives in sustainable building. One critical issue of water consumption is that in many areas, the demands on the supplying aquifer exceed its ability to replenish itself. To the maximum extent feasible, facilities should increase their dependence on water that is collected, used, purified, and reused on-site. The protection and conservation of water throughout the life of a building may be accomplished by designing for dual plumbing that recycles water in toilet flushing or by using water for washing of the cars. Waste-water may be minimized by utilizing water conserving fixtures such as ultra-low flush toilets and low-flow shower heads. Bidets help eliminate the use of toilet paper, reducing sewer traffic and increasing possibilities of re-using water on-site. Point of use water treatment and heating improves both water quality and energy efficiency while reducing the amount of water in circulation. The use of non-sewage and greywater for on-site use such as site-irrigation will minimize demands on the local aquifer.

Large commercial buildings with water and energy efficiency can qualify for an LEED Certification. Philadelphia's Comcast Center is the tallest building in Philadelphia. It's also one of the tallest buildings in the USA that is LEED Certified. Their environmental engineering consists of a hybrid central chilled water system which cools floor-by-floor with steam instead of water. Burn's Mechanical set-up the entire renovation of the 58 story, 1.4 million square foot sky scraper.

Materials Efficiency

Building materials typically considered to be 'green' include lumber from forests that have been certified to a third-party forest standard, rapidly renewable plant materials like bamboo and straw, dimension stone, recycled stone, recycled metal, and other products that are non-toxic, reusable, renewable, and/or recyclable. For concrete a high performance or Roman self-healing concrete is available. The EPA (Environmental Protection Agency) also suggests using recycled industrial goods, such as coal combustion products, foundry sand, and demolition debris in construction projects. Energy efficient building materials and appliances are promoted in the United States through energy rebate programs.

Indoor Environmental Quality Enhancement

The Indoor Environmental Quality (IEQ) category in LEED standards, one of the five environmental categories, was created to provide comfort, well-being, and productivity of occupants. The LEED IEQ category addresses design and construction guidelines especially: indoor air quality (IAQ), thermal quality, and lighting quality.

Indoor Air Quality seeks to reduce volatile organic compounds, or VOCs, and other air impurities such as microbial contaminants. Buildings rely on a properly designed ventilation system (passively/naturally or mechanically powered) to provide adequate ventilation of cleaner air from outdoors or recirculated, filtered air as well as isolated operations (kitchens, dry cleaners, etc.) from other occupancies. During the design and construction process choosing construction materials and interior finish products with zero or low VOC emissions will improve IAQ. Most building materials and cleaning/maintenance products emit gases, some of them toxic, such as many VOCs including formaldehyde. These gases can have a detrimental impact on occupants' health, comfort, and productivity. Avoiding these products will increase a building's IEQ. LEED, HQE and Green Star contain specifications on use of low-emitting interior. Draft LEED 2012 is about to expand the scope of the involved products. BREEAM limits formaldehyde emissions, no other VOCs. MAS Certified Green is a registered trademark to delineate low VOC-emitting products in the marketplace. The MAS Certified Green Program ensures that any potentially hazardous chemicals released from manufactured products have been thoroughly tested and meet rigorous standards established by independent toxicologists to address recognized long term health concerns. These IAQ standards have been adopted by and incorporated into the following programs: (1) The United States Green Building Council (USGBC) in their LEED rating system (2) The California Department of Public Health (CDPH) in their section 01350 standards (3) The Collaborative for High Performance Schools (CHPS) in their Best Practices Manual and (4) The Business and Institutional Furniture Manufacturers Association (BIFMA) in their level® sustainability standard.

Also important to indoor air quality is the control of moisture accumulation (dampness) leading to mold growth and the presence of bacteria and viruses as well as dust mites and other organisms and microbiological concerns. Water intrusion through a building's envelope or water condensing on cold surfaces on the building's interior can enhance and sustain microbial growth. A well-insulated and tightly sealed envelope will reduce moisture problems but adequate ventilation is also necessary to eliminate moisture from sources indoors including human metabolic processes, cooking, bathing, cleaning, and other activities.

Personal temperature and airflow control over the HVAC system coupled with a properly designed building envelope will also aid in increasing a building's thermal quality. Creating a high performance luminous environment through the careful integration of daylight and electrical light sources will improve on the lighting quality and energy performance of a structure.

Solid wood products, particularly flooring, are often specified in environments where occupants are known to have allergies to dust or other particulates. Wood itself is considered to be hypo-allergenic and its smooth surfaces prevent the buildup of particles common in soft finishes like carpet. The Asthma and Allergy Foundation of American recommends hardwood, vinyl, linoleum tile or slate flooring instead of carpet. The use of wood products can also improve air quality by absorbing or releasing moisture in the air to moderate humidity.

Interactions among all the indoor components and the occupants together form the processes that determine the indoor air quality. Extensive investigation of such processes is the subject of indoor air scientific research and is well documented in the journal Indoor Air.

Operations and Maintenance Optimization

No matter how sustainable a building may have been in its design and construction, it can only remain so if it is operated responsibly and maintained properly. Ensuring operations and maintenance(O&M) personnel are part of the project's planning and development process will help retain the green criteria designed at the onset of the project. Every aspect of green building is integrated into the O&M phase of a building's life. The addition of new green technologies also falls on the O&M staff. Although the goal of waste reduction may be applied during the design, construction and demolition phases of a building's life-cycle, it is in the O&M phase that green practices such as recycling and air quality enhancement take place. O&M staff should aim to establish best practices in energy efficiency, resource conservation, ecologically sensitive products and other sustainable practices. Education of building operators and occupants is key to effective implementation of sustainable strategies in O&M services.

Waste Reduction

Green architecture also seeks to reduce waste of energy, water and materials used during construction. For example, in California nearly 60% of the state's waste comes from commercial buildings During the construction phase, one goal should be to reduce the amount of material going to landfills. Well-designed buildings also help reduce the amount of waste generated by the occupants as well, by providing on-site solutions such as compost bins to reduce matter going to landfills.

To reduce the amount of wood that goes to landfill, Neutral Alliance (a coalition of government, NGOs and the forest industry) created the website dontwastewood.com. The site includes a variety of resources for regulators, municipalities, developers, contractors, owner/operators and individuals/homeowners looking for information on wood recycling.

When buildings reach the end of their useful life, they are typically demolished and hauled to landfills. Deconstruction is a method of harvesting what is commonly considered "waste" and reclaiming it into useful building material. Extending the useful life of a structure also reduces waste – building materials such as wood that are light and easy to work with make renovations easier.

To reduce the impact on wells or water treatment plants, several options exist. "Greywater", wastewater from sources such as dishwashing or washing machines, can be used for subsurface irrigation, or if treated, for non-potable purposes, e.g., to flush toilets and wash cars. Rainwater collectors are used for similar purposes.

Centralized wastewater treatment systems can be costly and use a lot of energy. An alternative to this process is converting waste and wastewater into fertilizer, which avoids these costs and shows other benefits. By collecting human waste at the source and running it to a semi-centralized biogas plant with other biological waste, liquid fertilizer can be produced. This concept was demonstrated by a settlement in Lubeck Germany in the late 1990s. Practices like these provide soil with organic nutrients and create carbon sinks that remove carbon dioxide from the atmosphere, offsetting greenhouse gas emission. Producing artificial fertilizer is also more costly in energy than this process.

Reduce Impact onto Electricity Network

Electricity networks are built based on peak demand (another name is peak load). Peak demand is measured in the units of watts (W). It shows how fast electrical energy is consumed. Residential electricity is often charged on electrical energy (kilowatt hour, kWh). Green buildings or sustainable buildings are often capable of saving electrical energy but not necessarily reducing peak demand.

When sustainable building features are designed, constructed and operated efficiently, peak demand can be reduced so that there is less desire for electricity network expansion and there is less impact onto carbon emission and climate change. These sustainable features can be good orientation, sufficient indoor thermal mass, good insulation, photovoltaic panels, thermal or electrical energy storage systems, smart building (home) energy management systems.

Cost and Payoff

The most criticized issue about constructing environmentally friendly buildings is the price. Photo-voltaics, new appliances, and modern technologies tend to cost more money. Most green buildings cost a premium of <2%, but yield 10 times as much over the entire life of the building. In regards to the financial benefits of green building, "Over 20 years, the financial payback typically exceeds the additional cost of greening by a factor of 4-6 times. And broader benefits, such as reductions in greenhouse gases (GHGs) and other pollutants have large positive impacts on surrounding communities and on the planet." The stigma is between the knowledge of up-front cost vs. life-cycle cost. The savings in money come from more efficient use of utilities which result in decreased energy bills. It is projected that different sectors could save $130 Billion on energy bills. Also, higher worker or student productivity can be factored into savings and cost deductions.

Numerous studies have shown the measurable benefit of green building initiatives on worker productivity. In general it has been found that, "there is a direct correlation between increased productivity and employees who love being in their work space." Specifically, worker productivity can be significantly impacted by certain aspects of green building design such as improved lighting, reduction of pollutants, advanced ventilation systems and the use of non-toxic building materials. In "The Business Case for Green Building", the U.S. Green Building Council gives another specific example of how commercial energy retrofits increase worker health and thus productivity, "People in the U.S. spend about 90% of their time indoors. EPA studies indicate indoor levels of pollutants may be up to ten times higher than outdoor levels. LEED-certified buildings are designed to have healthier, cleaner indoor environmental quality, which means health benefits for occupants."

Studies have shown over a 20-year life period, some green buildings have yielded $53 to $71 per square foot back on investment. Confirming the rentability of green building investments, further studies of the commercial real estate market have found that LEED and Energy Star certified buildings achieve significantly higher rents, sale prices and occupancy rates as well as lower capitalization rates potentially reflecting lower investment risk.

Regulation and Operation

As a result of the increased interest in green building concepts and practices, a number of organizations have developed standards, codes and rating systems that let government regulators, building professionals and consumers embrace green building with confidence. In some cases, codes are written so local governments can adopt them as bylaws to reduce the local environmental impact of buildings.

Green building rating systems such as BREEAM (United Kingdom), LEED (United States and Canada), DGNB (Germany), CASBEE (Japan), and VERDEGBCe (Spain) help consumers determine a structure's level of environmental performance. They award credits for optional building features that support green design in categories such as location and maintenance of building site, conservation of water, energy, and building materials, and occupant comfort and health. The number of credits generally determines the level of achievement.

Green building codes and standards, such as the International Code Council's draft International Green Construction Code, are sets of rules created by standards development organizations that establish minimum requirements for elements of green building such as materials or heating and cooling.

Some of the major building environmental assessment tools currently in use include:

- United States: International Green Construction Code (IGCC)

International Frameworks and Assessment Tools

IPCC Fourth Assessment Report

Climate Change 2007, the Fourth Assessment Report (AR4) of the United Nations Intergovernmental Panel on Climate Change (IPCC), is the fourth in a series of such reports. The IPCC was established by the World Meteorological Organization (WMO) and the United Nations Environment Programme (UNEP) to assess scientific, technical and socio-economic information concerning climate change, its potential effects and options for adaptation and mitigation.

UNEP and Climate Change

United Nations Environment Program UNEP works to facilitate the transition to low-carbon societies, support climate proofing efforts, improve understanding of climate change science, and raise public awareness about this global challenge.

GHG Indicator

The Greenhouse Gas Indicator: UNEP Guidelines for Calculating Greenhouse Gas Emissions for Businesses and Non-Commercial Organizations

Agenda 21

Agenda 21 is a programme run by the United Nations (UN) related to sustainable development. It is a comprehensive blueprint of action to be taken globally, nationally and locally by organizations of the UN, governments, and major groups in every area in which humans impact on the environment. The number 21 refers to the 21st century.

FIDIC's PSM

The International Federation of Consulting Engineers (FIDIC) Project Sustainability Management Guidelines were created in order to assist project engineers and other stakeholders in setting sustainable development goals for their projects that are recognized and accepted by as being in the interests of society as a whole. The process is also intended to allow the alignment of project goals with local conditions and priorities and to assist those involved in managing projects to measure and verify their progress.

The Project Sustainability Management Guidelines are structured with Themes and Sub-Themes under the three main sustainability headings of Social, Environmental and Economic. For each individual Sub-Theme a core project indicator is defined along with guidance as to the relevance of that issue in the context of an individual project.

The Sustainability Reporting Framework provides guidance for organizations to use as the basis for disclosure about their sustainability performance, and also provides stakeholders a universally applicable, comparable framework in which to understand disclosed information.

The Reporting Framework contains the core product of the Sustainability Reporting Guidelines, as well as Protocols and Sector Supplements. The Guidelines are used as the basis for all reporting.

They are the foundation upon which all other reporting guidance is based, and outline core content for reporting that is broadly relevant to all organizations regardless of size, sector, or location. The Guidelines contain principles and guidance as well as standard disclosures – including indicators – to outline a disclosure framework that organizations can voluntarily, flexibly, and incrementally, adopt.

Protocols underpin each indicator in the Guidelines and include definitions for key terms in the indicator, compilation methodologies, intended scope of the indicator, and other technical references.

Sector Supplements respond to the limits of a one-size-fits-all approach. Sector Supplements complement the use of the core Guidelines by capturing the unique set of sustainability issues faced by different sectors such as mining, automotive, banking, public agencies and others.

IPD Environment Code

The IPD Environment Code was launched in February 2008. The Code is intended as a good practice global standard for measuring the environmental performance of corporate buildings. Its aim is to accurately measure and manage the environmental impacts of corporate buildings and enable property executives to generate high quality, comparable performance information about their buildings anywhere in the world. The Code covers a wide range of building types (from offices to airports) and aims to inform and support the following;

- Creating an environmental strategy
- Inputting to real estate strategy
- Communicating a commitment to environmental improvement
- Creating performance targets
- Environmental improvement plans
- Performance assessment and measurement
- Life cycle assessments
- Acquisition and disposal of buildings
- Supplier management
- Information systems and data population
- Compliance with regulations
- Team and personal objectives

IPD estimate that it will take approximately three years to gather significant data to develop a robust set of baseline data that could be used across a typical corporate estate.

ISO 21931

ISO/TS 21931:2006, Sustainability in building construction—Framework for methods of assessment for environmental performance of construction works—Part 1: Buildings, is intended to provide a general framework for improving the quality and comparability of methods for assessing the environmental performance of buildings. It identifies and describes issues to be taken into account when using methods for the assessment of environmental performance for new or existing building properties in the design, construction, operation, refurbishment and deconstruction stages. It is not an assessment system in itself but is intended be used in conjunction with, and following the principles set out in, the ISO 14000 series of standards

References

- "TAIPEI 101 Receives LEED v4 Certification from U.S. Green Building Council as All-Time Top Scorer". Taipei 101. Taipei Financial Center Corp. Retrieved 7 November 2016
- Mao, X., Lu, H., & Li, Q. (2009). International Conference on Management and Service Science, 2009. MASS '09., 1-5. doi:10.1109/ICMSS.2009.5303546
- Agarwal, Yuvraj; Balaji, Bharathan; Gupta, Rajesh; Lyles, Jacob; Wei, Michael; Weng, Thomas. "Occupancy-Driven Energy Management for Smart Building Automation" (PDF). BuildSys. Retrieved 11 November 2016
- Yan Ji and Stellios Plainiotis (2006): Design for Sustainability. Beijing: China Architecture and Building Press. ISBN 7-112-08390-7
- Pushkar, S., Becker, R., & Katz, A.(2005). A methodology for design of environmentally optimal buildings by variable grouping. Building and Environment, 40. doi:10.1016/j.buildenv.2004.09.004
- Fuerst, Franz; McAllister, Pat. An Investigation of the Effect of Eco-Labeling on Office Occupancy Rates. 2009. [6] Retrieved: November 5, 2010
- Lee YS, Guerin DA, Indoor environmental quality differences between office types in LEED-certified buildings in the US, Building and Environment (2009), doi:10.1016/j.buildenv.2009.10.019

6

An Integrated Study of Green Economics

Green economy mainly concentrates on decreasing environmental risks by aiming at a sustainable environment. Industries have started to adopt these practices to promote green economy. Eco-capitalism, ecological economics and deep ecology are some important topics related to the subject matter. Green politics is best understood in confluence with the major topics listed in the following chapter.

Green Economy

The green economy is defined as an economy that aims at reducing environmental risks and ecological scarcities, and that aims for sustainable development without degrading the environment. It is closely related with ecological economics, but has a more politically applied focus. The 2011 UNEP Green Economy Report argues "that to be green, an economy must not only be efficient, but also fair. Fairness implies recognising global and country level equity dimensions, particularly in assuring a just transition to an economy that is low-carbon, resource efficient, and socially inclusive."

A feature distinguishing it from prior economic regimes is the direct valuation of natural capital and ecological services as having economic value and a full cost accounting regime in which costs externalized onto society via ecosystems are reliably traced back to, and accounted for as liabilities of, the entity that does the harm or neglects an asset.

Green Sticker and ecolabel practices have emerged as consumer facing measurements of friendliness to the environment and sustainable development. Many industries are starting to adopt these standards as a viable way to promote their greening practices in a globalizing economy.

"Green" Economists and Economics

"Green economics" is loosely defined as any theory of economics by which an economy is considered to be component of the ecosystem in which it resides (after Lynn Margulis). A holistic approach to the subject is typical, such that economic ideas are commingled with any number of other subjects, depending on the particular theorist. Proponents of feminism, postmodernism, the ecology movement, peace movement, Green politics, green anarchism and anti-globalization movement have used the term to describe very different ideas, all external to some equally ill-defined "mainstream" economics.

The use of the term is further ambiguated by the political distinction of Green parties which are formally organized and claim the capital-G "Green" term as a unique and distinguishing mark. It is thus preferable to refer to a loose school of "'green economists"' who generally advocate shifts towards a green economy, biomimicry and a fuller accounting for biodiversity.

Some economists view green economics as a branch or subfield of more established schools. For instance, it is regarded as classical economics where the traditional land is generalized to natural capital and has some attributes in common with labor and physical capital (since natural capital assets like rivers directly substitute for man-made ones such as canals). Or, it is viewed as Marxist economics with nature represented as a form of Lumpenproletariat, an exploited base of non-human workers providing surplus value to the human economy, or as a branch of neoclassical economics in which the price of life for developing vs. developed nations is held steady at a ratio reflecting a balance of power and that of non-human life is very low.

An increasing commitment by the UNEP (and national governments such as the UK) to the ideas of natural capital and full cost accounting under the banner 'green economy' could blur distinctions between the schools and redefine them all as variations of "green economics". As of 2010 the Bretton Woods institutions (notably the World Bank and International Monetary Fund (via its "Green Fund" initiative) responsible for global monetary policy have stated a clear intention to move towards biodiversity valuation and a more official and universal biodiversity finance. Taking these into account targeting not less but radically zero emission and waste is what is promoted by the Zero Emissions Research and Initiatives. The UNEP 2011 Green Economy Report informs that "[b]ased on existing studies, the annual financing demand to green the global economy was estimated to be in the range US$ 1.05 to US$ 2.59 trillion. To place this demand in perspective, it is about one-tenth of total global investment per year, as measured by global Gross Capital Formation."

Definition

Karl Burkart defines a green economy as based on six main sectors:

- Renewable energy
- Green buildings
- Sustainable transport
- Water management
- Waste management
- Land management

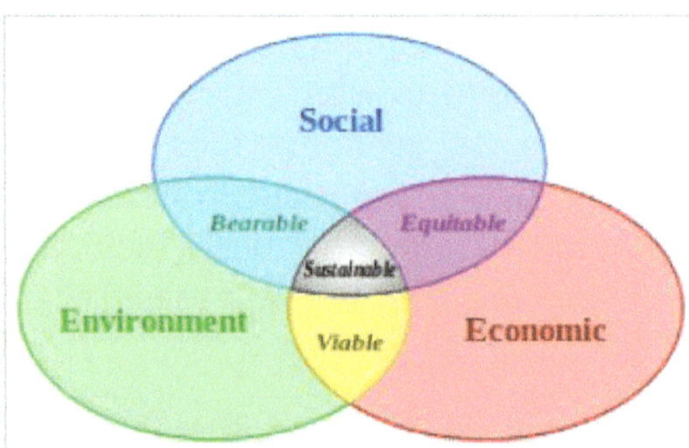

The three pillars of sustainability.

The International Chamber of Commerce (ICC) representing global business defines green economy as "an economy in which economic growth and environmental responsibility work together in a mutually reinforcing fashion while supporting progress on social development".

In 2012, the ICC published the Green Economy Roadmap, containing contributions from experts from around the globe brought together in a two-year consultation process. The Roadmap represents a comprehensive and multidisciplinary effort to clarify and frame the concept of "green economy". It highlights the essential role of business in bringing solutions to common global challenges. It sets out the following 10 conditions which relate to business/intra-industry and collaborative action for a transition towards a green economy:

- Open and competitive markets
- Metrics, accounting, and reporting
- Finance and investment
- Awareness
- Life cycle approach
- Resource efficiency and decoupling
- Employment
- Education and skills
- Governance and partnership
- Integrated policy and decision-making

Ecological Measurements

Measuring economic output and progress is done through the use of economic index indicators. Green indices emerged from the need to measure human ecological impact, efficiency sectors like transport, energy, buildings and tourism, as well as the investment flows targeted to areas like renewable energy and cleantech innovation.

1. 2010 - 2016 Global Green Economy Index™ (GGEI), published by consultancy Dual Citizen LLC is in its 5th edition. It measures the green economic performance and perceptions of it in 80 countries and 50 cities along four main dimensions of leadership & climate change, efficiency sectors, markets & investment and the environment.
2. 2009 - 2012 Green City Index A global study commissioned by Siemens
3. 2009 - 2013 Circles of Sustainability project scored 5 cities in 5 separate countries.

Ecological footprint measurements are a way to gauge anthropogenic impact and are another standard used by municipal governments.

Green Energy Issues

Green economies require green energy generation based on renewable energy to replace fossil fuels as well as energy conservation and efficient energy use.

There is justification for market failure to respond to environmental protection and climate protection needs with the excuse that high external costs and high initial costs for research, development, and marketing of green energy sources and green products prevents firms from voluntarily reducing their ecological footprints. The green economy may need government subsidies as market incentives to motivate firms to invest and produce green products and services. The German Renewable Energy Act, legislations of many other member states of the European Union and the American Recovery and Reinvestment Act of 2009, all provide such market incentives. However, other experts argue that green strategies can be highly profitable for corporations that understand the business case for sustainability and can market green products and services beyond the traditional green consumer.

Criticisms

A number of organisations and individuals have criticised aspects of the 'Green Economy', particularly the mainstream conceptions of it based on using price mechanisms to protect nature, arguing that this will extend corporate control into new areas from forestry to water. The research organisation ETC Group argues that the corporate emphasis on bio-economy "will spur even greater convergence of corporate power and unleash the most massive resource grab in more than 500 years." Venezuelan professor Edgardo Lander says that the UNEP's report, *Towards a Green Economy*, while well-intentioned "ignores the fact that the capacity of existing political systems to establish regulations and restrictions to the free operation of the markets – even when a large majority of the population call for them – is seriously limited by the political and financial power of the corporations." Ulrich Hoffmann, in a paper for UNCTAD also says that the focus on Green Economy and "green growth" in particular, "based on an evolutionary (and often reductionist) approach will not be sufficient to cope with the complexities of climate change" and "may rather give much false hope and excuses to do nothing really fundamental that can bring about a U-turn of global greenhouse gas emissions. Clive Spash, an ecological economist, has criticised the use of economic growth to address environmental losses, and argued that the Green Economy, as advocated by the UN, is not a new approach at all and is actually a diversion from the real drivers of environmental crisis. He has also criticised the UN's project on the economics of ecosystems and biodiversity (TEEB), and the basis for valuing ecosystems services in monetary terms.

Eco-capitalism

Eco-capitalism, also known as environmental capitalism or green capitalism, is the view that capital exists in nature as "natural capital" (ecosystems that have ecological yield) on which all wealth depends, and therefore, market-based government policy instruments (such as cap and trade systems) should be used to resolve environmental problems.

The term "Blue Greens" is often applied to those who espouse eco-capitalism. It is considered as the right-wing equivalent to Red Greens.

History

The roots of eco-capitalism can be traced back to the late 1960s. The "Tragedy of the Commons", an essay published in 1968 in Science by Garrett Hardin, claimed the inevitability of malthusian catastrophe due to liberal or democratic government's policies to leave family size matters to the family, and enabling the welfare state to willingly care for potential overpopulation. Hardin argued that if families were given freedom of choice in the matter, but were removed from a welfare state, parents choosing to overbear would not have the resources to provide for their "litter", thus solving the problem of overpopulation. This represents an early argument made from an eco-capitalist standpoint: overpopulation would technically be solved by a free market. John Baden, a collaborator with Garrett Hardin on other works including *Managing the Commons,* founded the Political Economy Research Center (now called the Property and Environment Research Center) in 1982. As one of the first eco-capitalist organizations created, PERC's ongoing mission is "improving environmental quality through property rights and markets". The most popular eco-capitalist idea was emissions trading, or more commonly, cap and trade. Emissions trading, a market-based approach that allows polluting entities to purchase or be allocated permits, began being researched in the late 1960s. International emissions trading was significantly popularized in the 1990s when the United Nations adopted the Kyoto Protocol in 1997.

Eco-capitalist Theorists

Terry L. Anderson is a graduate of the University of Montana, and received his Ph.D from Washington University. Anderson is currently serving as the co-chair of the Hoover Institution's Property Rights, Freedom and Prosperity task force. Anderson advocates that free markets can be both economically beneficial and environmentally protective. Anderson specializes in how markets impact Native American communities and their economies.

Bruce Yandle, a graduate of Mercer University, attended Georgia State University where he earned a MBA and PhD. Yandle is currently serving as dean emeritus of Clemson University's college of business. Yandle is prominent in the field of eco-capitalism for his story of the "Bootlegger and the Baptist". Yandle's theory of the Bootlegger and the Baptist posits that ethical groups, religious institutions and business captains can align their organizations in the interest of regulation and economic growth.

Paul Hawken

Paul Hawken decided at a young age to dedicate his life to making business eco-friendlier. Hawken's is the architect of the United States first natural foods company, Erewhon Trading Company where all products were organically composed. Hawken's continued to make an impact on the business world by founding the research organization, Natural Capital Institute, and developed, Wiser Earth, a program focused on providing a platform for all to communicate about the environment. Not only has Paul Hawken set a good example for how to transform economy into eco-capitalism, but also has authored hundreds of publications, including four best selling books. In his writings, Hawken stresses that many smart ecological options are out there for businesses that will help save the environment, while also continuing to bring profit. One idea discussed in his book, *Natural Capitalism: Creating the Next Industrial Revolution*, is the possibility of developing lightweight, electricity-powered cars as an alternative to our current transportation issue. Hawken attributes the hesitancy of adopting these options to lack of knowledge of these alternatives and high initial costs. Paul Hawken is now the head of OneSun, Inc., an energy corporation concentrated on low-cost solar.

Lester Brown

Lester Brown began his career as a tomato farmer in New Jersey, until earning his degree at Rutgers University and traveling to a rural India for a six-month study of the country's food and population crisis. From this point on, Brown's focus was mostly on finding alternatives that would solve the world's population and resources problem. With financial support from Rockefeller Brothers Fund, Brown created the Worldwatch Institute, the first dedicated to researching global environmental problems. In 2001, Brown found the Earth Policy Institute, an organization that outlined a vision for creating an environmentally sustainable economy. Over the course of his career, Lester Brown has authored over 50 books and received 25 honorary degrees. In his publications, Brown discusses how the key to an eco-friendly economy is an honest market and replacing harmful aspects of the environment, like fossil fuels with renewable energy. On June 2015, Lester Brown retired from Earth Policy and closed the institute.

Transition to Eco-capitalism

The ideology of eco-capitalism was adopted to satisfy two competing needs: 1) the desire for generating profit by businesses in a capitalist society and 2) the urgency for proper actions to address a struggling environment negatively impacted by human activity. Under the doctrine of eco-capitalism, businesses commodify the act of addressing environmental issues.

The following are common principles in the transition to eco-capitalism.

Green Consumption

At the conception of the ideology, major theorists of eco-capitalism, Paul Hawken, Lester Brown, and Francis Cairncross, saw an opportunity to establish a different approach to environmentalism in a capitalist society. These theorists thought that not only producers but also consumers could shoulder the social responsibility of environmental restoration if "green technology, green taxes, green labeling, and eco-conscious shopping" existed. The resulting "shopping our way to sustainability" mentality encouraged the development of organic farming, renewable energy, green certifications as well as other eco-friendly practices.

Carbon Trading

Creating perhaps the first major eco-capitalist endorsement, many political and economic institutions support a system of pollution credits. Such a system, which assigns property rights to emissions, is considered to be the most "efficient and effective" way for regulating greenhouse gas emissions in the current neoliberal global economy. Especially in the case of tradable pollution credits, the resulting market-based system of emissions regulation is believed to motivate businesses to invest in technology that reduce greenhouse gas emissions using positive reinforcement (i.e. ability to trade unused credits) and punishment (i.e. the need to buy more credits).

Full Cost Accounting

Environmental full cost accounting explains corporate actions on the basis of the triple bottom line, which is best summarized as "people, planet, and profit". As a concept of corporate social responsibility, full cost accounting not only considers social and economic costs and benefits but also the environmental implications of specific corporate actions.

Genuine Progress Indicator

The current standard of using the gross domestic product (GDP) as an indicator of welfare is criticized for being inaccurate. An alternative to GDP, the genuine progress indicator compensates for the shortcomings of the GDP as a welfare indicator by accounting for environmental harms as well as other factors that affect consumption, such as crime and income inequality.

Criticisms

Majority of the criticism from traditionally unregulated capitalism is due to eco-capitalism's increased regulation. Pollution credits (as a means for regulating greenhouse gas emissions) is traditionally at odds with economically conservative ideologies. Elements of unregulated capitalism prefer environmental issues to be addressed by individuals who may allocate their own income and wealth, oppose the commodification of by-products like carbon emissions, and emphasize positive incentives to maintain resources through free-market competition and entrepreneurship.

Proponents of eco-capitalism view environmental reform like pollution credits as a more transformative and progressive system. According to these Proponents, since free market capitalism as inherently expansionist in tendency, ignoring environmental responsibility is a danger to the environment. Approximately 36% of Americans are deeply concerned about climate issues. Proponents of Eco Capitalism typically favor political environmentalism, which emphasizes negative incentives like regulation and taxes to encourage the conservation of resources and prevent environmental harm.

Political theorist, Antonio Gramsci, cites theories of common sense, which suggests that, in general, free market capitalism absent of environmental reform, is ingrained in the minds of its members as the only viable and successful form of economic organization through cultural hegemony. Therefore, the proposal of any alternate economic system, like eco-capitalism, must overcome the predominant common sense and economic status quo in order to develop opposing theories. Nonetheless, movements in the United States and abroad have continued to push for reforms to protect the environment in current capitalistic systems.

Another political theorist, Daniel Tanuro, explains, in his book "Green Capitalism: Why it Can't Work", that for green capitalism to be successful, it would have to replace current mainstream capitalism with Eco-socialist methods, while defying corporate interests:

- "If by 'green capitalism' we understand a system in which the qualitative, social and ecological parameters are taken in account by the numerous competing capitals, that is to say even within economic activity as an endogenous mechanism, then we are completely deluded. In fact, we would be talking about a form of capitalism in which the law of value was no longer in operation, which is a contradiction in terms"(112)

However, Tanuro adds that social and economical change to the current capitalist systems is necessary, because technology will invariably increase emissions as manufacturing processes and distribution systems progress. Tanuro argues for changes in three areas:

1. Use of transportation methods
2. Agriculture and dietary changes
3. Overall consumer lifestyle and market spending

Despite this argument, critics still claim that green consumption, sustainable behavior on the part of the consumer, is not enough to be instituted as a socio-environmental solution. In accordance with hegemony, capitalism agrees that the government has little control over market and buyers, sellers, and consumers ultimately drive the market. In contrast, in green capitalism, the government would have more control therefore; consumers do not have direct power over the market, and should not be held accountable. Thus, going against the established monetary system of capitalism in the U.S. and spread throughout the globally.

Environmental Scholar Bill McKibben proposes "full scale climate mobilization" to address environmental decay. During World War II, vehicle manufacturers and general goods manufacturers shifted to producing weapons, military vehicles and war time goods. McKibben argues that, to combat environmental change, the American Military Industrial Complex and other national arms producers could shift to producing solar panels, wind turbines and other environmental products in an Eco-Capitalist system.

Scholar Elliot Sperber counters McKibben's argument, citing that industrial environmental mobilization favoring eco-capitalism would exacerbate socioeconomic stratification. Sperber counters the notion that "full scale climate mobilization" and the production it implies is the best immediate solution for addressing climate change. Because Eco Capitalism is still capitalistic, it relies on production of goods. Sperber argues for the production of fewer goods (i.e. fewer plastics, fewer vehicles) to minimize carbon footprints.

Apparent criticisms have risen concerns and need for social and economical transformation on both ends of the political and theoretical divide. Nonetheless, they have shaped the way the majority public has viewed and contributed to capitalism and continue to both actively change the innate structure of the economic system and enhance it for further economic stability.

Appeal of Renewable Energy in the Capitalist Market

Tom Randall, a correspondent specializing in renewable energy for Bloomberg, calls to attention that wind and solar (energy sources) are "outperforming" fossil fuels. In terms of investments, clean energy outperforms both gas and coal by a 2-1 margin. This positive margin may be attributed to the consistently falling price of renewable energy production. Renewable energy sources hold assertive advantages over fossil fuels because they exist as technologies, not fuels. As time proceeds, renewable energy becomes inevitably more efficient as technology adapts. Technologies for extracting fuels may change, but the fuels remain as constants. Both the solar and wind industries have proven growth over time: Over the last 15 years, the solar industry has doubled seven times and the wind industry has doubled four times. In contrast, the fossil fuel industry has declined over the last 15 years. America's coal industry has lost 75 percent of its value within the past few years.

Renewable energy sources also gain advantages over the fossil fuel industry through international governmental support. Globally, governments implement subsidies to boost the renewable energy industry. Concurrently, various global efforts fight against fossil fuel production and use. The demand for renewable energy sources has skyrocketed in the last 15 years, while fossil fuels have drastically fallen in demand (in capitalist societies).

The worldwide concern of climate change (also known as global warming) is notably the largest contributor to the green energy industry's rapid acceleration, just as it is largely responsible for the decline of the fossil fuel industry. The overwhelming scientific consensus of climate change's reality and its potential catastrophic effects have caused a large part of the world's population to respond with panic and immediate action. While the world's response has been strong, environmentalists and climate scientists do not believe the response has been strong enough to counter climate change's effects, and that the transition from fossil fuels to renewable energy sources is moving far too slowly.

The global efforts and concerns of both governments and individuals to take action regarding implementing and transforming a society's energy sources from fossil fuels to renewable energy sources show the enormous potential of the green energy market. This potential is seen in the countless renewable energy projects under way. Currently, there are over 4,000 major solar projects being implemented. These, and all renewable energy projects, set goals of long-term economic benefit.

The Global Apollo Programme, set up by both economists and scientists, has a goal of creating a solar capability that can stand as a cheaper alternative to coal-fueled power plants by 2025. In capitalist markets, solar energy has the very real potential of becoming a direct competitor to coal plants in less than a decade.

Ecological Economics

Ecological economics/eco-economics (also called Ecolonomy or Bioeconomics of Georgescu-Roegen) refers to both a transdisciplinary and interdisciplinary field of academic research addressing the interdependence and coevolution of human economies and natural ecosystems, both intertemporally and spatially. By treating the economy as a subsystem of Earth's larger ecosystem, and by emphasizing the preservation of natural capital, the field of ecological economics is differentiated from environmental economics, which is the mainstream economic analysis of the environment. One survey of German economists found that ecological and environmental economics are different schools of economic thought, with ecological economists emphasizing strong sustainability and rejecting the proposition that natural capital can be substituted by human-made capital.

Ecological economics was founded in the 1980s as a modern discipline on the works of and interactions between various European and American academics. The related field of green economics is, in general, a more politically applied form of the subject.

According to ecological economist Malte Faber, ecological economics is defined by its focus on nature, justice, and time. Issues of intergenerational equity, irreversibility of environmental change, uncertainty of long-term outcomes, and sustainable development guide ecological economic analysis and valuation. Ecological economists have questioned fundamental mainstream economic approaches such as cost-benefit analysis, and the separability of economic values from scientific research, contending that economics is unavoidably normative rather than positive (i.e. descriptive). Positional analysis, which attempts to incorporate time and justice issues, is proposed as an alternative. Ecological economics shares many of its perspectives with feminist economics, including the focus on sustainability, nature, justice and care values.

History and Development

The first principles, deriving from the radiochemist FA Soddy, were laid out in his 1926 book "Wealth, Money (Virtual Wealth) and Debt" in 1926. Early modern interest in ecology and economics dates back to the 1940s in the work of K. William Kapp and Karl Polanyi and the 1960s in work by Kenneth Boulding and Herman Daly. However, the first organized meetings of modern ecological economists occurred in the 1980s. These began in 1982, at the instigation of Lois Banner, with a meeting held in Sweden (including Robert Costanza, Herman Daly, Charles Hall, Bruce Hannon, H.T. Odum, and David Pimentel). Most were ecosystem ecologists or mainstream environmental economists, with the exception of Daly. In 1987, Daly and Costanza edited an issue of *Ecological Modeling* to test the waters. A book entitled *Ecological Economics*, by Juan Martinez-Alier, was published later that year. 1989 saw the foundation of the International Society for Ecological Economics and publication of its journal, *Ecological Economics*, by Elsevier. Robert Costanza was the first president of the society and first editor of the journal, currently edited by Richard Howarth.

European conceptual founders include Nicholas Georgescu-Roegen (1971), K. William Kapp (1950) and Karl Polanyi (1944). Some key concepts of what is now ecological economics are evident in the writings of E.F. Schumacher, whose book *Small Is Beautiful – A Study of Economics as if People Mattered* (1973) was published just a few years before the first edition of Herman Daly's comprehensive and persuasive *Steady-State Economics* (1977). Other figures include ecologists C.S. Holling, H.T. Odum and Robert Costanza, biologist Gretchen Daily and physicist Robert Ayres. CUNY geography professor David Harvey explicitly added ecological concerns to political economic literature. This parallel development in political economy has been continued by analysts such as sociologist John Bellamy Foster.

The antecedents can be traced back to the Romantics of the 19th century as well as some Enlightenment political economists of that era. Concerns over population were expressed by Thomas Malthus, while John Stuart Mill predicted the desirability of the *stationary state* of an economy. Mill thereby anticipated later insights of modern ecological economists, but without having had their experience of the social and ecological costs of the Post–World War II economic expansion. As Martinez-Alier explores in his book the debate on energy in economic systems can also be traced into the 19th century e.g. Nobel prize-winning chemist, Frederick Soddy (1877–1956). Soddy criticized the prevailing belief of the economy as a perpetual motion machine, capable of generating infinite wealth—a criticism echoed by his intellectual heirs in the now emergent field of ecological economics.

The Romanian economist Nicholas Georgescu-Roegen (1906–1994), who was among Daly's teachers at Vanderbilt University, provided ecological economics with a modern conceptual framework based on the material and energy flows of economic production and consumption. His *magnum opus*, *The Entropy Law and the Economic Process* (1971), has been highly influential.

Articles by Inge Ropke (2004, 2005) and Clive Spash (1999) cover the development and modern history of ecological economics and explain its differentiation from resource and environmental economics, as well as some of the controversy between American and European schools of thought. An article by Robert Costanza, David Stern, Lining He, and Chunbo Ma responded to a call by Mick Common to determine the foundational literature of ecological economics by using citation analysis to examine which books and articles have had the most influence on the development of the field. However, citations analysis has itself proven controversial and similar work has been criticized by Clive Spash for attempting to pre-determine what is regarded as influential in ecological economics through study design and data manipulation. In addition, the journal Ecological Economics has itself been criticized for swamping the field with mainstream economics.

Nature and Ecology

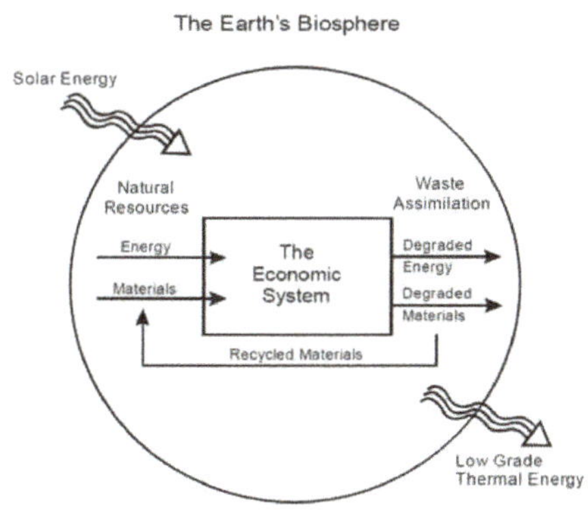

Natural resources flow through the economy and end up as waste and pollution

A simple circular flow of income diagram is replaced in ecological economics by a more complex flow diagram reflecting the input of solar energy, which sustains natural inputs and environmental services which are then used as units of production. Once consumed, natural inputs pass out of the economy as pollution and waste. The potential of an environment to provide services and materials is referred to as an "environment's source function", and this function is depleted as resources are consumed or pollution contaminates the resources. The "sink function" describes an environment's ability to absorb and render harmless waste and pollution: when waste output exceeds the limit of the sink function, long-term damage occurs. Some persistent pollutants, such as some organic pollutants and nuclear waste are absorbed very slowly or not at all; ecological economists emphasize minimizing "cumulative pollutants". Pollutants affect human health and the health of the climate.

The economic value of natural capital and ecosystem services is accepted by mainstream environmental economics, but is emphasized as especially important in ecological economics. Ecological economists may begin by estimating how to maintain a stable environment before assessing the cost in dollar terms. Ecological economist Robert Costanza led an attempted valuation of the global ecosystem in 1997. Initially published in *Nature*, the article concluded on $33 trillion with a range from $16 trillion to $54 trillion (in 1997, total global GDP was $27 trillion). Half of the value went to nutrient cycling. The open oceans, continental shelves, and estuaries had the highest total value, and the highest per-hectare values went to estuaries, swamps/floodplains, and seagrass/

algae beds. The work was criticized by articles in *Ecological Economics* Volume 25, Issue 1, but the critics acknowledged the positive potential for economic valuation of the global ecosystem.

The Earth's carrying capacity is a central issue in ecological economics. Early economists such as Thomas Malthus pointed out the finite carrying capacity of the earth, which was also central to the MIT study *Limits to Growth*. Diminishing returns suggest that productivity increases will slow if major technological progress is not made. Food production may become a problem, as erosion, an impending water crisis, and soil salinity (from irrigation) reduce the productivity of agriculture. Ecological economists argue that industrial agriculture, which exacerbates these problems, is not sustainable agriculture, and are generally inclined favorably to organic farming, which also reduces the output of carbon.

Global wild fisheries are believed to have peaked and begun a decline, with valuable habitat such as estuaries in critical condition. The aquaculture or farming of piscivorous fish, like salmon, does not help solve the problem because they need to be fed products from other fish. Studies have shown that salmon farming has major negative impacts on wild salmon, as well as the forage fish that need to be caught to feed them.

Since animals are higher on the trophic level, they are less efficient sources of food energy. Reduced consumption of meat would reduce the demand for food, but as nations develop, they tend to adopt high-meat diets similar to that of the United States. Genetically modified food (GMF) a conventional solution to the problem, presents numerous problems – Bt corn produces its own Bacillus thuringiensis toxin/protein, but the pest resistance is believed to be only a matter of time.[:31] The overall effect of GMF on yields is contentious, with the USDA and FAO acknowledging that GMFs do not necessarily have higher yields and may even have reduced yields.

Global warming is now widely acknowledged as a major issue, with all national scientific academies expressing agreement on the importance of the issue. As the population growth intensifies and energy demand increases, the world faces an energy crisis. Some economists and scientists forecast a global ecological crisis if energy use is not contained – the Stern report is an example. The disagreement has sparked a vigorous debate on issue of discounting and intergenerational equity.

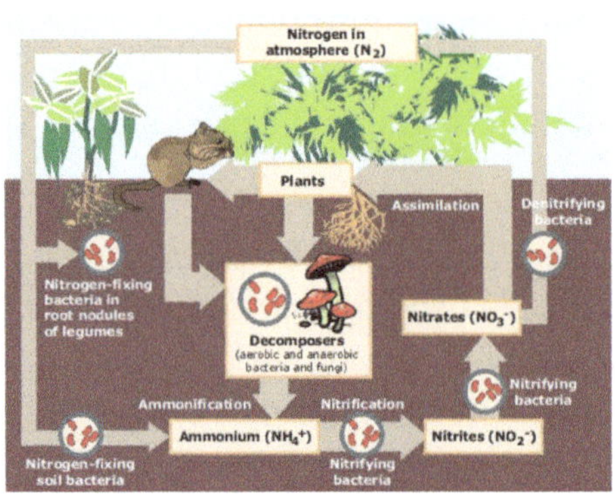

Nitrogen cycle

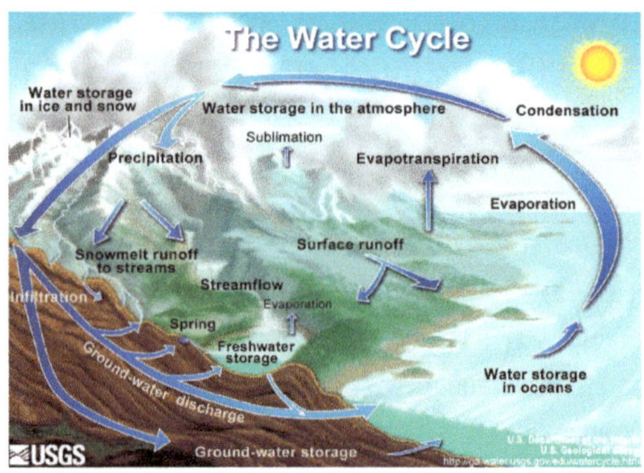

Water cycle

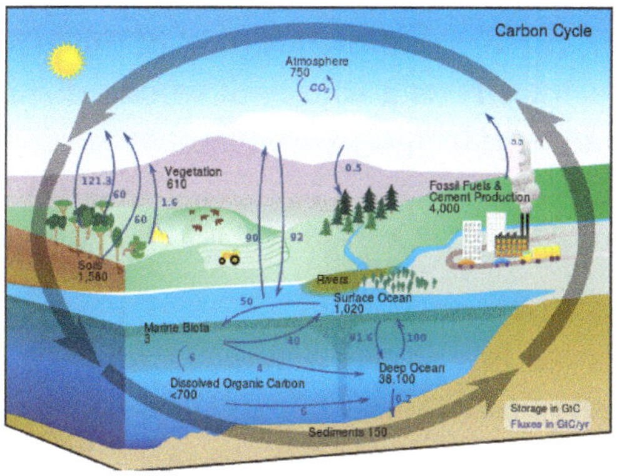

Carbon cycle

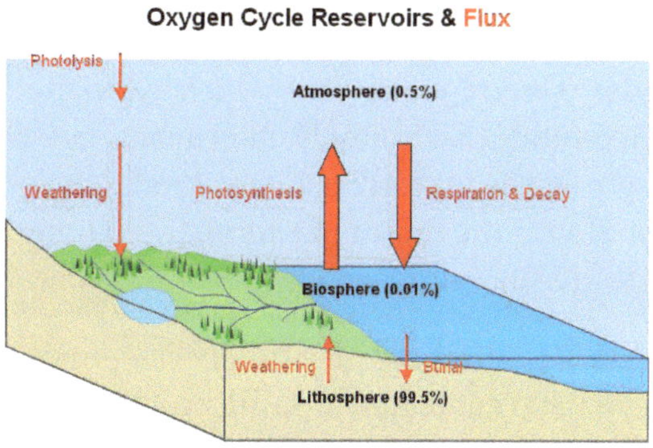

Oxygen cycle

Ethics

Mainstream economics has attempted to become a value-free 'hard science', but ecological economists argue that value-free economics is generally not realistic. Ecological economics is more willing to entertain alternative conceptions of utility, efficiency, and cost-benefits such as positional analysis or multi-criteria analysis. Ecological economics is typically viewed as economics for sustainable development, and may have goals similar to green politics.

Schools of Thought

Various competing schools of thought exist in the field. Some are close to resource and environmental economics while others are far more heterodox in outlook. An example of the latter is the *European Society for Ecological Economics*. An example of the former is the Swedish *Beijer International Institute of Ecological Economics*. Clive Spash has argued for the classification of the ecological economics movement, and more generally work by different economic schools on the environment, into three main categories. These are the mainstream new resource economists, the new environmental pragmatists, and the more radical social ecological economists. International survey work comparing the relevance of the categories for mainstream and heterodox economists shows some clear divisions between environmental and ecological economists.

Differentiation from Mainstream Schools

Some ecological economists prioritise adding natural capital to the typical capital asset analysis of land, labor, and financial capital. These ecological economists then use tools from mathematical economics as in mainstream economics, but may apply them more closely to the natural world. Whereas mainstream economists tend to be technological optimists, ecological economists are inclined to be technological sceptics. They reason that the natural world has a limited carrying capacity and that its resources may run out. Since destruction of important environmental resources could be practically irreversible and catastrophic, ecological economists are inclined to justify cautionary measures based on the precautionary principle.

The most cogent example of how the different theories treat similar assets is tropical rainforest ecosystems, most obviously the Yasuni region of Ecuador. While this area has substantial deposits of bitumen it is also one of the most diverse ecosystems on Earth and some estimates establish it has over 200 undiscovered medical substances in its genomes - most of which would be destroyed by logging the forest or mining the bitumen. Effectively, the instructional capital of the genomes is undervalued by analyses which view the rainforest primarily as a source of wood, oil/tar and perhaps food. Increasingly the carbon credit for leaving the extremely carbon-intensive ("dirty") bitumen in the ground is also valued - the government of Ecuador set a price of US$350M for an oil lease with the intent of selling it to someone committed to never exercising it at all and instead preserving the rainforest.

While this natural capital and ecosystems services approach has proven popular amongst many it has also been contested as failing to address the underlying problems with mainstream economics, growth, market capitalism and monetary valuation of the environment. Critiques concern the need to create a more meaningful relationship with Nature and the non-human world than evident in the instrumentalism of shallow ecology and the environmental economists commodification of everything external to the market system.

Topics

Among the topics addressed by ecological economics are methodology, allocation of resources, weak versus strong sustainability, energy economics, energy accounting and balance, environmental services, cost shifting, and modeling.

Methodology

A primary objective of ecological economics (EE) is to ground economic thinking and practice in physical reality, especially in the laws of physics (particularly the laws of thermodynamics) and in knowledge of biological systems. It accepts as a goal the improvement of human well-being through development, and seeks to ensure achievement of this through planning for the sustainable development of ecosystems and societies. Of course the terms development and sustainable development are far from lacking controversy. Richard B. Norgaard argues traditional economics has hi-jacked the development terminology in his book *Development Betrayed*.

Well-being in ecological economics is also differentiated from welfare as found in mainstream economics and the 'new welfare economics' from the 1930s which informs resource and environmental economics. This entails a limited preference utilitarian conception of value i.e., Nature is valuable to our economies, that is because people will pay for its services such as clean air, clean water, encounters with wilderness, etc.

Ecological economics is distinguishable from neoclassical economics primarily by its assertion that the economy is embedded within an environmental system. Ecology deals with the energy and matter transactions of life and the Earth, and the human economy is by definition contained within this system. Ecological economists argue that neoclassical economics has ignored the environment, at best considering it to be a subset of the human economy.

The neoclassical view ignores much of what the natural sciences have taught us about the contributions of nature to the creation of wealth e.g., the planetary endowment of scarce matter and energy, along with the complex and biologically diverse ecosystems that provide goods and ecosystem services directly to human communities: micro- and macro-climate regulation, water recycling, water purification, storm water regulation, waste absorption, food and medicine production, pollination, protection from solar and cosmic radiation, the view of a starry night sky, etc.

There has then been a move to regard such things as natural capital and ecosystems functions as goods and services. However, this is far from uncontroversial within ecology or ecological economics due to the potential for narrowing down values to those found in mainstream economics and the danger of merely regarding Nature as a commodity. This has been referred to as ecologists 'selling out on Nature'. There is then a concern that ecological economics has failed to learn from the extensive literature in environmental ethics about how to structure a plural value system.

Allocation of Resources

Resource and neoclassical economics focus primarily on the efficient allocation of resources, and less on two other fundamental economic problems which are central to ecological economics: distribution (equity) and the scale of the economy relative to the ecosystems upon which it is reliant. Ecological Economics also makes a clear distinction between growth (quantitative increase in economic output) and development (qualitative improvement of the quality of life) while arguing that neoclassical economics confuses the two. Ecological economists point out that, beyond modest levels, increased per-capita consumption (the typical economic measure of "standard of living") does not necessarily lead to improvement in human well-being, while this same consumption can have harmful effects on the environment and broader societal well-being.

Weak Versus Strong Sustainability

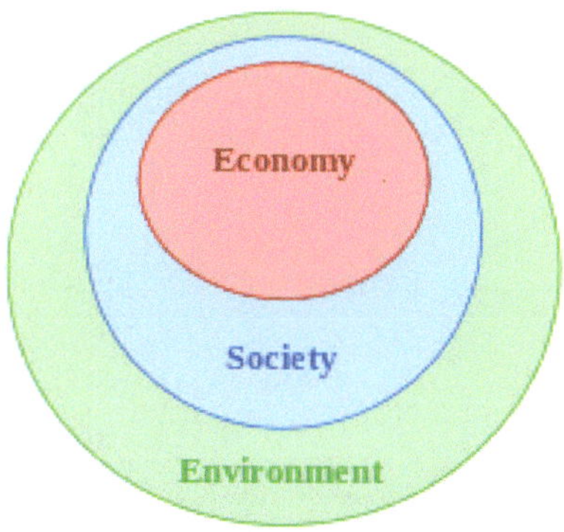

The three nested systems of sustainability - the economy wholly contained by society, wholly contained by the biophysical environment. Clickable.

Ecological economics challenges the conventional approach towards natural resources, claiming that it undervalues natural capital by considering it as interchangeable with human-made capital—labor and technology.

The impending depletion of natural resources and increase of climate-changing greenhouse gasses should motivate us to examine how political, economic and social policies can benefit from alternative energy. Interestingly enough, shifting our dependence on fossil fuels with specific interest within just one of the above-mentioned factors easily benefits at least one other. For instance, photo voltaic (or solar) panels have a 15% efficiency when absorbing the sun's energy, but its construction demand has increased 120% within both commercial and residential properties. Additionally, this construction has led to a roughly 30% increase in work demands (Chen).

The potential for the substitution of man-made capital for natural capital is an important debate in ecological economics and the economics of sustainability. There is a continuum of views among economists between the strongly neoclassical positions of Robert Solow and Martin Weitzman, at one extreme and the 'entropy pessimists', notably Nicholas Georgescu-Roegen and Herman Daly, at the other.

Neoclassical economists tend to maintain that man-made capital can, in principle, replace all types of natural capital. This is known as the weak sustainability view, essentially that every technology can be improved upon or replaced by innovation, and that there is a substitute for any and all scarce materials.

At the other extreme, the strong sustainability view argues that the stock of natural resources and ecological functions are irreplaceable. From the premises of strong sustainability, it follows that economic policy has a fiduciary responsibility to the greater ecological world, and that sustainable development must therefore take a different approach to valuing natural resources and ecological functions.

Recently, Stanislav Shmelev developed a new methodology for the assessment of progress at the macro scale based on multi-criteria methods, which allows consideration of different perspectives, including strong and weak sustainability or conservationists vs industrialists and aims to search for a 'middle way' by providing a strong neo-Keynsian economic push without putting excessive pressure on the natural resources, including water or producing emissions, both directly and indirectly.

Energy Economics

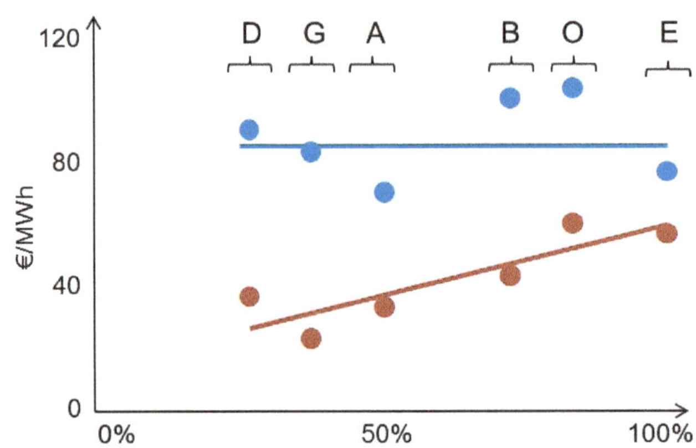

Exergy analysis can be performed to find connections between economic value and the physical world. Here the costs of heating (vertical axis) are compared with the exergy content of different energy carriers (horizontal axis). Red dots and trend line indicate energy prices for consumers, blue dots and trend line indicate total price for consumers including capital expenditure for the heating system. Energy carriers included are district heating (D), ground-source heat pump (G), exhaust air heat pump (A), bioenergy meaning firewood (B), heating oil (O) and direct electric heating (E).

A key concept of energy economics is net energy gain, which recognizes that all energy requires energy to produce. To be useful the energy return on energy invested (*EROEI*) has to be greater than one. The net energy gain from production coal, oil and gas has declined over time as the easiest to produce sources have been most heavily depleted.

Ecological economics generally rejects the view of energy economics that growth in the energy supply is related directly to well being, focusing instead on biodiversity and creativity - or natural capital and individual capital, in the terminology sometimes adopted to describe these economically. In practice, ecological economics focuses primarily on the key issues of uneconomic growth and quality of life. Ecological economists are inclined to acknowledge that much of what is important in human well-being is not analyzable from a strictly economic standpoint and suggests an interdisciplinary approach combining social and natural sciences as a means to address this.

Thermoeconomics is based on the proposition that the role of energy in biological evolution should be defined and understood through the second law of thermodynamics, but also in terms of such economic criteria as productivity, efficiency, and especially the costs and benefits (or profitability) of the various mechanisms for capturing and utilizing available energy to build biomass and do work. As a result, thermoeconomics is often discussed in the field of ecological economics, which itself is related to the fields of sustainability and sustainable development.

Exergy analysis is performed in the field of industrial ecology to use energy more efficiently. The term *exergy*, was coined by Zoran Rant in 1956, but the concept was developed by J. Willard Gibbs. In recent decades, utilization of exergy has spread outside of physics and engineering to the fields of industrial ecology, ecological economics, systems ecology, and energetics.

Energy Accounting and Balance

An energy balance can be used to track energy through a system, and is a very useful tool for determining resource use and environmental impacts, using the First and Second laws of thermodynamics, to determine how much energy is needed at each point in a system, and in what form that energy is a cost in various environmental issues. The energy accounting system keeps track of energy in, energy out, and non-useful energy versus work done, and transformations within the system.

Scientists have written and speculated on different aspects of energy accounting.

Environmental Services

A study was carried out by Costanza and colleagues to determine the 'price' of the services provided by the environment. This was determined by averaging values obtained from a range of studies conducted in very specific context and then transferring these without regard to that context. Dollar figures were averaged to a per hectare number for different types of ecosystem e.g. wetlands, oceans. A total was then produced which came out at 33 trillion US dollars (1997 values), more than twice the total GDP of the world at the time of the study. This study was criticized by pre-ecological and even some environmental economists - for being inconsistent with assumptions of financial capital valuation - and ecological economists - for being inconsistent with an ecological economics focus on biological and physical indicators.

The whole idea of treating ecosystems as goods and services to be valued in monetary terms remains controversial. A common objection is that life is precious or priceless, but this demonstrably degrades to it being worthless under the assumptions of any branch of economics. Reducing human bodies to financial values is a necessary part of every branch of economics and not always in the direct terms of insurance or wages. Economics, in principle, assumes that conflict is reduced by agreeing on voluntary contractual relations and prices instead of simply fighting or coercing or tricking others into providing goods or services. In doing so, a provider agrees to surrender time and take bodily risks and other (reputation, financial) risks. Ecosystems are no different from other bodies economically except insofar as they are far less replaceable than typical labour or commodities.

Despite these issues, many ecologists and conservation biologists are pursuing ecosystem valuation. Biodiversity measures in particular appear to be the most promising way to reconcile financial and ecological values, and there are many active efforts in this regard. The growing field of biodiversity finance began to emerge in 2008 in response to many specific proposals such as the Ecuadoran Yasuni proposal or similar ones in the Congo. US news outlets treated the stories as a "threat" to "drill a park" reflecting a previously dominant view that NGOs and governments had the primary responsibility to protect ecosystems. However Peter Barnes and other commentators have recently argued that a guardianship/trustee/commons model is far more effective and takes the decisions out of the political realm.

Commodification of other ecological relations as in carbon credit and direct payments to farmers to preserve ecosystem services are likewise examples that enable private parties to play more direct roles protecting biodiversity. The United Nations Food and Agriculture Organization achieved near-universal agreement in 2008 that such payments directly valuing ecosystem preservation and encouraging permaculture were the only practical way out of a food crisis. The holdouts were all English-speaking countries that export GMOs and promote "free trade" agreements that facilitate their own control of the world transport network: The US, UK, Canada and Australia.

Not 'externalities', but Cost Shifting

Ecological economics is founded upon the view that the neoclassical economics (NCE) assumption that environmental and community costs and benefits are mutually canceling *"externalities"* is not warranted. Joan Martinez Alier, for instance shows that the bulk of consumers are automatically excluded from having an impact upon the prices of commodities, as these consumers are future generations who have not been born yet. The assumptions behind future discounting, which assume that future goods will be cheaper than present goods, has been criticized by Fred Pearce and by the recent Stern Report (although the Stern report itself does employ discounting and has been criticized for this and other reasons by ecological economists such as Clive Spash).

Concerning these externalities, some like the eco-businessman Paul Hawken argue an orthodox economic line that the only reason why goods produced unsustainably are usually cheaper than goods produced sustainably is due to a hidden subsidy, paid by the non-monetized human environment, community or future generations. These arguments are developed further by Hawken, Amory and Hunter Lovins to promote their vision of an environmental capitalist utopia in *Natural Capitalism: Creating the Next Industrial Revolution.*

In contrast, ecological economists, like Joan Martinez-Alier, appeal to a different line of reasoning. Rather than assuming some (new) form of capitalism is the best way forward, an older ecological economic critique questions the very idea of internalizing externalities as providing some corrective to the current system. The work by Karl William Kapp explains why the concept of "externality" is a misnomer. In fact the modern business enterprise operates on the basis of shifting costs onto others as normal practice to make profits. Charles Einsentein has argued that this method of privatising profits while socialising the costs through externalities, passing the costs to the community, to the natural environment or to future generations is inherently destructive As social ecological economist Clive Spash has noted, externality theory fallaciously assumes environmental and social problems are minor aberrations in an otherwise perfectly functioning efficient economic system. Internalizing the odd externality does nothing to address the structural systemic problem and fails to recognize the all pervasive nature of these supposed 'externalities'.

Ecological-economic Modeling

Mathematical modeling is a powerful tool that is used in ecological economic analysis. Various approaches and techniques include: evolutionary, input-output, neo-Austrian modeling, entropy and thermodynamic models, multi-criteria, and agent-based modeling, the environmental Kuznets curve, and Stock-Flow consistent model frameworks. System dynamics and GIS are techniques applied, among other, to spatial dynamic landscape simulation modeling. The Matrix accounting methods of Christian Felber provide a more sophisticated method for identifying "the common good"

Criticism

Assigning monetary value to natural resources such as biodiversity, and the emergent ecosystem services is often viewed as a key process in influencing economic practices, policy, and decision-making. While this idea is becoming more and more accepted among ecologists and conservationist, some argue that it is inherently false.

McCauley argues that ecological economics and the resulting ecosystem service based conservation can be harmful. He describes four main problems with this approach:

Firstly, it seems to be assumed that all ecosystem services are financially beneficial. This is undermined by a basic characteristic of ecosystems: they do not act specifically in favour of any single species. While certain services might be very useful to us, such as coastal protection from hurricanes by mangroves for example, others might cause financial or personal harm, such as wolves hunting cattle. The complexity of Eco-systems makes it challenging to weigh up the value of a given species. Wolves play a critical role in regulating prey populations, the absence of such an apex predator in the Scottish Highlands have caused the over population of deer, preventing afforestation, which increases the risk of flooding and damage to property.

Secondly, allocating monetary value to nature would make its conservation reliant on markets that fluctuate. This can lead to devaluation of services that were previously considered financially beneficial. Such is the case of the bees in a forest near former coffee plantations in Finca Santa Fe, Costa Rica. The pollination services were valued to over US$60,000 a year, but soon after the study, coffee prices dropped and the fields were replanted with pineapple. Pineapple does not require bees to be pollinated, so the value of their service dropped to zero.

Thirdly, conservation programmes for the sake of financial benefit underestimate human ingenuity to invent and replace ecosystem services by artificial means. McCauley argues that such proposals are deemed to have a short lifespan as the history of technology is about how Humanity developed artificial alternatives to nature's services and with time passing the cost of such services tend to decrease. This would also lead to the devaluation of ecosystem services.

Lastly, it should not be assumed that conserving ecosystems is always financially beneficial as opposed to alteration. In the case of the introduction of the Nile perch to Lake Victoria, the ecological consequence was decimation of native fauna. However, this same event is praised by the local communities as they gain significant financial benefits from trading the fish.

McCauley argues that, for these reasons, trying to convince decision-makers to conserve nature for monetary reasons is not the path to be followed, and instead appealing to morality is the ultimate way to campaign for the protection of nature.

Deep Ecology

Deep ecology is an ecological and environmental philosophy promoting the inherent worth of living beings regardless of their instrumental utility to human needs, plus a radical restructuring of modern human societies in accordance with such ideas. Deep ecology supports the views that the natural world is a harmony of homeostasis due to the complex inter-relationships in which the life of organisms is actively regulated, partially by the existence of other organisms within the biosphere, in order to constantly be conducive to life. These views are scientifically encouraged by Living systems theory and the Gaia hypothesis. Human interference with or destruction of the natural world poses a threat therefore not only to humans but to all organisms of our biosphere.

Deep ecology's core principle is the belief that the living environment as a whole should be respected and regarded as having certain inalienable legal rights to live and flourish, independent of its utilitarian instrumental benefits for human use. Deep ecology is often framed in terms of the idea of a much broader sociality; it recognizes diverse communities of life on Earth that are composed not only through biotic factors but also, where applicable, through ethical relations, that is, the valuing of other beings as more than just resources. It describes itself as "deep" because it regards itself as looking more deeply into the actual reality of humanity's relationship with the natural world arriving at philosophically more profound conclusions than that of the prevailing view of ecology as a branch of biology. The movement does not subscribe to anthropocentric environmentalism (which is concerned with conservation of the environment only for exploitation by and for human purposes) since deep ecology is grounded in a quite different set of philosophical assumptions. Deep ecology takes a more holistic view of the world human beings live in and seeks to apply to life the understanding that the separate parts of the ecosystem (including humans) function as a whole. This philosophy provides a foundation for the environmental, ecology, and green movements and has fostered a new system of environmental ethics advocating wilderness preservation, human population control, and simple living.

Origins

In his original 1972/73 deep ecology paper, Arne Næss claims the deep ecology movement arose from scientists – ecologists who were out in the field studying the biodiversity and wild ecosystems throughout the world. They were also doing the work of philosophers, laying the foundations for the Age of Ecology and a new ecological worldview to replace the anthropocentric, mastery of Nature, and modernist worldview arising in the 17th and 18th centuries. Three of the most influential ecological spokespersons of the 1960s were Rachel Carson, David Brower, and Paul Ehrlich. Some consider the publication of Rachel Carson's book Silent Spring (1962) as the beginning of the contemporary, long-range deep ecology movement. When her book appeared there was a long-standing movement for conservation of land and resources, as well as support for creating parks and other areas devoted to preserving wilderness and spectacular nature. Carson's writings were especially influential because they clearly showed how human well-being depends on the condition of whole biotic communities. She explained in practical terms how living beings are interrelated within ecosystems. She explained how pesticides used to control mosquitoes and other insects led to declines in some bird populations. "Silent Spring" helped show how complex food webs and networks of biotic relationships function. Since humans are at the top of many food chains, exposure to chemicals becomes more concentrated as these move up the chains. The chemicals also can be stored in human tissues and gradually accumulate over time, adversely affecting health. Carson showed the need for deep changes in human practices and ways of living.

The 1960s was a decade of vigorous social activism in the United States, Canada, Western Europe, and Australia. Some activism focused on war and peace and the issue of nuclear weapons. A well-known early environmental organization started with a focus on nuclear tests and their environmental hazards. Some people in British Columbia, Canada, were opposed to the test of a nuclear weapon by the US government on Amchitka Island. They hired a fishing vessel and sailed towards the nuclear test site in protest. This action led to the founding of "Greenpeace", which became more identified with environmental issues as time went by. These great movements were further catalyzed by the now iconic images of the whole Earth floating in space taken during the return of the Apollo space missions from their journey to the moon. Among the astronauts that witnessed seeing the whole Earth firsthand was Edgar D. Mitchell, who in 1971, during the return mission of Apollo 14, had an epiphany that what is needed to solve the eco-crisis "is a transformation of consciousness".

Principles

Proponents of deep ecology believe that the world does not exist as a resource to be freely exploited by humans. If material goods do not guarantee happiness beyond a very moderate level, and over-consumption is endangering the biosphere, defining a new non-consumptive paradigm of well-being seems primordial, such a paradigm would be non-acquisitive/non-consumerist and non-hierarchical in relation to our place on Earth. The ethics of deep ecology hold that the survival of any part is dependent upon the well-being of the whole. Proponents of deep ecology offer an eight-tier platform to elucidate their claims:

1. The well-being and flourishing of human and nonhuman life on Earth have value in themselves (synonyms: intrinsic value, inherent value). These values are independent of the usefulness of the nonhuman world for human purposes.

2. Richness and diversity of life forms contribute to the realization of these values and are also values in themselves.

3. Humans have no right to reduce this richness and diversity except to satisfy vital human needs.

4. The flourishing of human life and cultures is compatible with a substantial decrease of the human population. The flourishing of nonhuman life requires such a decrease.

5. Present human interference with the nonhuman world is excessive, and the situation is rapidly worsening.

6. Policies must therefore be changed. These policies affect basic economic, technological, and ideological structures. The resulting state of affairs will be deeply different from the present.

7. The ideological change is mainly that of appreciating life quality (dwelling in situations of inherent value) rather than adhering to an increasingly higher standard of living. There will be a profound awareness of the difference between big and great.

8. Those who subscribe to the foregoing points have an obligation directly or indirectly to try to implement the necessary changes.

These principles can be reduced to three simple propositions:

1. Wilderness and biodiversity preservation;

2. Human population control;

3. Simple living (or treading lightly on the planet).

Development

The phrase "deep ecology" was coined by the Norwegian philosopher Arne Næss in 1973. Næss rejected the idea that beings can be ranked according to their relative value. For example, judgments on whether an animal has an eternal soul, whether it uses reason or whether it has consciousness (or indeed higher consciousness) have all been used to justify the ranking of the human animal as superior to other animals. Næss states that from an ecological point of view "the right of all forms [of life] to live is a universal right which cannot be quantified. No single species of living being has more of this particular right to live and unfold than any other species."

This metaphysical idea is elucidated in Warwick Fox's claim that humanity and all other beings are "aspects of a single unfolding reality". As such deep ecology would support the view of Aldo Leopold in his book *A Sand County Almanac* that humans are "plain members of the biotic community". They also would support Leopold's land ethic: "a thing is right when it tends to preserve the integrity, stability and beauty of the biotic community. It is wrong when it tends otherwise." Daniel Quinn, in his novel *Ishmael*, showed that an anthropocentric myth underlies our current view of the world.

The ecological problems faced by the world today are partly due to the loss of traditional knowledge, values, and ethics of behavior that celebrate the intrinsic value and sacredness of the natural world and that give the preservation of Nature prime importance. Correspondingly, the assumption of human superiority to other life forms, as if we were granted royalty status over Nature; the idea that Nature is mainly here to serve human will and purpose. Deep ecology had a response to the anthropocentric view and several different actors played a role in its development.

Prominent among them is Joseph W. Meeker, who in 1973 told George Sessions about Arne Næss, whom Meeker knew personally. As Warwick Fox related, "One of the things that initially interested Sessions about Næss was Næss's strong interest in, and innovative approach to, the work of Spinoza. Sessions says that he had himself 'arrived at Spinoza as the answer to the process of teaching history of philosophy by about 1972 and independently of being in contact with Næss.'" Sessions therefore wrote to Næss at this time, beginning a lifelong association. Meeker's (1972, 1997) book *The Comedy of Survival: Studies in Literary Ecology* emerged through the work of scholars seeking an environmental ethic. That book represents Meeker's founding work in literary ecology and ecocriticism, which demonstrates the relationship between the literary arts and scientific ecology, especially humankind's consideration of comedy and tragedy. It reminds readers that adaptive behaviors (comedy) promote survival, whereas tragedy estranges from other life forms. This thesis rests on Meeker's study of comparative literature, his work with biologist Konrad Lorenz, and his work as a field ecologist in the National Park service in Alaska, Oregon, and California.

Deep ecology offers a philosophical basis for environmental advocacy which may, in turn, guide human activity against perceived self-destruction. Deep ecology and environmentalism hold that the science of ecology shows that ecosystems can absorb only limited change by humans or other dissonant influences. Further, both hold that the actions of modern civilization threaten global ecological well-being. Ecologists have described change and stability in ecological systems in various ways, including homeostasis, dynamic equilibrium, and "flux of nature". Regardless of which model is most accurate, environmentalists contend that massive human economic activity has pushed the biosphere far from its "natural" state through reduction of biodiversity, climate change, and other influences. As a consequence, civilization is causing mass extinction at a rate between 100 species a day and possibly 140,000 species a year, which is 10,000 times the background rate of extinction. Deep ecologists hope to influence social and political change through their philosophy. Næss has proposed, as Nicholas Goodrick-Clarke writes, "that the earth's human population should be reduced to about 100 million."

Environmental Education

Ecology in the narrow sense refers to the biological science of ecology. However, ecological paradigms and principles are being developed and applied in almost all disciplines, and these paradigms have to do with the way we approach understanding the relationships and inter-connections within and between living beings which give to each its special place and identity. Human ecology, e.g., must certainly take account of the role of our subjective lives and spiritual needs, as well as our biological ones, in terms of their ecological effects. Ecology in this sense is not a reductionist undertaking, but a movement toward a more whole (or holistic) vision and understanding of world processes. Deep ecology seeks to look into all levels of existence and might be viewed as radical by some; for them a more anthropocentric view is appropriated because it put humans at the center. Learning how to live in harmony with our surroundings is beneficial because stopping the global extinction crisis and achieving true ecological sustainability will require rethinking our values as a society. In that way education seems to be the best way to start. Sustainability education aims to help learners understand their interconnectedness with all life, to become creative problem solvers and active citizens, and to engage personally and intellectually in shaping our common future. Experiential learning and critical pedagogy are central to providing opportunities for learners to engage in transformative sustainability learning. The "Environment" broadly defined, remains somewhat neglected within development studies, despite a substantial increase in contributions to the field over the fifteen years since 2000. Undergraduate and postgraduate courses (with some notable exceptions) often "add on" environmental issues as special lectures or modules, and there remains a tendency for those who are grounded in the material and discursive struggles that define the discipline to consider the environment as an exotic special interest, a problem that manifests itself in societies that have the leisure to care about the natural world. Development of a modern education model promoting patriotism and civic responsibility, active social position and healthy lifestyle is closely linked to the development of environmental responsibility in the younger generation. Development of environmentally responsible personality in individual is of particular importance for graduates of educational institutions. Environmental education could be integrated in different curriculum in most fields: education for sustainable development in the context of ecopedagogy.

Ecopedagogy calls for the remaking of capitalist practices and seeks to re-engage democracy to include multispecies interests in the face of our current global ecological crisis. It does so by using different ideas that challenge the way we see education. In Critical Pedagogy, Ecoliteracy and Planetary Crisis: The Ecopedagogy Movement, Richard Kahn (2010) reformulates Herbert Marcuse's critical theories of society, and supports the kind of education that seizes the power of radical environmental activists and supports the earth democracy in which multispecies interests are represented. The destruction of habitats and threats to biodiversity resulting from expansion of human population and consumption is rarely addressed in a way that confronts students with the necessity to consider moral implications of such destruction. Pedagogically, a return to education associated with significant life experiences, such as hiking in wilderness areas as a youth; as well as strategically significant education, action competence, social learning, and variations and combinations of those and many other pedagogical approaches developed in the past 40 years. Some of these pedagogical approaches have been disputed—for example, the belief that experiencing environment first hand is an essential component of engaging people in conservation has been disputed by arguments that these education efforts have been informed by behaviorist socio-psychology models that assumed a linear causality between education experience and pro-environmental behavior. Rather, the critics have argued that people's environmental behaviors are too complex and contextually dependent to be captured by a simple casual model. The process of environmental education of schoolchildren has the following methodological characteristics:

- Goal-setting as the projected results reflects a model of environmentally responsible personality, taking into account trends in the development of key elements of education system; all natural sciences are involved in the development of basic ecological concepts.
- The introduction of interactive training methods takes place at the high school level in teaching self-reflection, hypothesizing, predicting; school natural science education is rebuilt on the basis of system approach in accordance with the planned ecologization results. Implementation of relevant methodology will promote successful development of environmentally responsible personality in high school graduates.

In higher education, the analysis of students' individual writing assignments after viewing films/documentaries presents an interesting case of using radical 'messages' within the aims of environmental education in order to trigger both student's engagement and critical thinking. The case study "If a Tree Falls and Everybody Hears the Sound" provides an example of how environmental advocacy and the objective of pluralistic education can be combined as mutually supportive means of achieving both democratic learning in which students' individual opinions are seen as extremely valuable, and simultaneously provide an example of the type of ecopedagogy that supports learning for environmental sustainability. The role of environmental advocacy can be crucially important if the interests of all planetary citizens—and not just one species—are to be taken seriously.

Sources

Scientific

Old-growth forest in Biogradska Gora National Park, Montenegro

Næss and Fox do not claim to use logic or induction to derive the philosophy directly from scientific ecology but rather hold that scientific ecology directly implies the metaphysics of deep ecology, including its ideas about the self and further, that deep ecology finds scientific underpinnings in the fields of ecology and system dynamics.

In their 1985 book *Deep Ecology*, Bill Devall and George Sessions describe a series of sources of deep ecology. They include the science of ecology itself, and cite its major contribution as the rediscovery in a modern context that "everything is connected to everything else." They point out that some ecologists and natural historians, in addition to their scientific viewpoint, have developed a deep ecological consciousness—for some a political consciousness and at times a spiritual consciousness. This is a perspective beyond the strictly human viewpoint, beyond anthropocentrism. Among the scientists they mention specifically are Rachel Carson, Aldo Leopold, John Livingston, Paul R. Ehrlich and Barry Commoner, together with Frank Fraser Darling, Charles Sutherland Elton, Eugene Odum and Paul Sears.

A further scientific source for deep ecology adduced by Devall and Sessions is the "new physics", which they describe as shattering Descartes's and Newton's vision of the universe as a machine explainable in terms of simple linear cause and effect. They propose that Nature is in a state of constant flux and reject the idea of observers as existing independent of their environment. They refer to Fritjof Capra's *The Tao of Physics* and *The Turning Point* for their characterisation of how the new physics leads to metaphysical and ecological views of interrelatedness, which, according to Capra, should make deep ecology a framework for future human societies. Devall and Sessions also credit the American poet and social critic Gary Snyder—with his devotion to Buddhism, Native American studies, the outdoors, and alternative social movements—as a major voice of wisdom in the evolution of their ideas.

The Gaia hypothesis was also an influence on the development of deep ecology.

Spiritual

The central spiritual tenet of deep ecology is that the human species is a part of the Earth, not separate from it, and as such human existence is dependent on the diverse organisms within the natural world each playing a role in the natural economy of the biosphere. Coming to an awareness of this reality involves a transformation of an outlook that presupposes humanity's superiority over the natural world. This self-realisation or "re-earthing" is used for an individual to intuitively gain an ecocentric perspective. The notion is based on the idea that the more we *expand the self* to identify with "others" (people, animals, ecosystems), the more we realize ourselves. Transpersonal psychology has been used by Warwick Fox to support this idea. Deep ecology has influenced the development of contemporary Ecospirituality.

A number of spiritual and philosophical traditions including Native American, Buddhist and Jain are drawn upon in a continuing critique of the philosophical assumptions of the modern European mind which has enabled and led to what is seen as an increasingly unsustainable level of disregard to towards the rights and needs of the natural world and its ability to continue to support human life. In relation to the Judeo-Christian tradition, Næss offers the following criticism: "The arrogance of stewardship [as found in the Bible] consists in the idea of superiority which underlies the thought that we exist to watch over nature like a highly respected middleman between the Creator and Creation." This theme had been expounded in Lynn Townsend White, Jr.'s 1967 article "The Historical Roots of Our Ecological Crisis", in which however he also offered as an alternative Christian view of man's relation to nature that of Saint Francis of Assisi, who he says spoke for the equality of all creatures, in place of the idea of man's domination over creation. Næss' further criticizes the reformation's view of creation as property to be put into maximum productive use: a view used frequently in the past to exploit and dispossess native populations. Many Protestant sects today regard the Bible's call for man to have stewardship of the earth as a call for the care for creation, rather than for exploitation.

The original Christian teachings on property support the Franciscan/stewardship interpretation of the Bible. Against this view, Martin Luther condemned church ownership of lands because "they did not want to use that property in an economically productive fashion. At best they used it to produce prayers. Luther, and other Reformation leaders insisted that it should be used, not to relieve men from the necessity of working, but as a tool for making more goods. The attitude of the Reformation was practically, "not prayers, but production." And production, not for consumption, but for more production." This justification was offered to support secular takings of church endowments and properties.

Philosophical Roots

Spinoza

Arne Næss, who first wrote about the idea of deep ecology, from the early days of developing this outlook conceived Baruch Spinoza as a philosophical source.

Others have followed Naess' inquiry, including Eccy de Jonge, in *Spinoza and Deep Ecology: Challenging Traditional Approaches to Environmentalism*, and Brenden MacDonald, in *Spinoza, Deep Ecology, and Human Diversity—Realization of Eco-Literacies.*

One of the topical centres of inquiry connecting Spinoza to Deep Ecology is "self-realization." Arne Næss in *The Shallow and the Deep, Long-Range Ecology movement* and *Spinoza and the Deep Ecology Movement* for discussion on the role of Spinoza's conception of self-realization and its link to deep ecology.

Criticism, Debate, and Response

Knowledge of Non-human Interests

Animal rights activists state that for an entity to require rights and protection intrinsically, it must have interests. Deep ecology is criticised for assuming that living things such as plants, for example, have their own interests as they are manifested by the plant's behavior—for instance, self-preservation being considered an expression of a will to live. Deep ecologists claim to *identify* with non-human nature, and in doing so, deny those who claim that non-human (or non-sentient) lifeforms' needs or interests are nonexistent or unknowable. The criticism is that the interests that a deep ecologist attributes to non-human organisms such as survival, reproduction, growth, and prosperity are really human interests. This is sometimes construed as a pathetic fallacy or anthropomorphism, in which "the earth is endowed with 'wisdom', wilderness equates with 'freedom', and life forms are said to emit 'moral' qualities."

"Deepness"

Deep ecology is criticised for its claim to being *deeper* than alternative theories, which by implication are *shallow*. When Arne Næss coined the term *deep ecology*, he compared it favourably with *shallow environmentalism* which he criticized for its utilitarian and anthropocentric attitude to nature and for its materialist and consumer-oriented outlook. Against this is Arne Næss's own view that the "depth" of deep ecology resides in the persistence of its penetrative questioning, particularly in asking "Why?" when faced with initial answers.

Writer William D. Grey believes that developing a non-anthropocentric set of values is "a hopeless quest". He seeks an improved "shallow" view, writing, "What's wrong with shallow views is not their concern about the well-being of humans, but that they do not really consider enough in what that well-being consists. We need to develop an enriched, fortified anthropocentric notion of human interest to replace the dominant short-term, sectional and self-regarding conception."

Bookchin's Criticisms

Some critics, particularly social ecologist Murray Bookchin, have interpreted deep ecology as being hateful toward humanity, due in part to the characterization of humanity by some deep ecologists, such as David Foreman of Earth First!, as a pathological infestation on the Earth. Bookchin therefore asserts that "deep ecology, formulated largely by privileged male white academics, has managed to bring sincere naturalists like Paul Shepard into the same company as patently antihumanist and macho mountain men like David Foreman who preach a gospel that humanity is some kind of cancer in the world of life." Bookchin mentions that some, like Foreman, defend seemingly anti-human measures, such as severe population control and the claim regarding the Third World that "the best thing would be to just let nature seek its own balance, to let the people there just starve". However, Bookchin himself later admitted that "statements made by Earth First! activists are not to be confused with those made by deep ecology theorists". Ecophilosopher Warwick Fox similarly "warns critics not to commit the fallacy of 'misplaced misanthropy.' That is, just because deep ecology criticizes an arrogant anthropocentrism does *not* mean that deep ecology is misanthropic." Likewise, *The Deep Ecology Movement: An Introductory Anthology* attempts to clarify that "deep ecologists have been the strongest critics of anthropocentrism, so much so that they have often been accused of a mean-spirited misanthropy"; however, "deep ecology is actually vitally concerned with humans realizing their best potential" and "is explicit in offering a vision of an alternative way of living that is joyous and enlivening."

Bookchin's second major criticism is that deep ecology fails to link environmental crises with authoritarianism and hierarchy. Social ecologists like him believe that environmental problems are firmly rooted in the manner of human social interaction, and suggest that deep ecologists fail to recognise the potential for human beings to solve environmental issues through a change of cultural attitudes. According to Bookchin, it is a social reconstruction alone that "can spare the biosphere from virtual destruction." Though some deep ecologists may reject the argument that ecological behavior is rooted in the social paradigm (which, according to their view, would be an anthropocentric fallacy), others in fact embrace this argument, such as the adherents to the deep ecologist movement Deep Green Resistance.

Botkin's Criticism

Daniel Botkin has likened deep ecology to its antithesis, the wise use movement, when he says that they both "misunderstand scientific information and then arrive at conclusions based on their misunderstanding, which are in turn used as justification for their ideologies. Both begin with an ideology and are political and social in focus." Elsewhere, though, he asserts that deep ecology must be taken seriously in the debate about the relationship between humans and nature because it challenges the fundamental assumptions of Western philosophy. Botkin has also criticized Næss's restatement and reliance upon the balance of nature idea and the perceived contradiction between his argument that all species are morally equal and his disparaging description of pioneering species.

Response

Some writers have misunderstood Naess, taking his Ecosophy T, with its Self-realization norm, as something meant to characterize the whole deep ecology movement as part of a single philosophy called "deep ecology." Naess was not doing either of these. He emphasized that movements cannot be precisely defined, but only roughly characterized by very general statements. They are often united internationally by means of such principles as found in the United Nations (UN) Earth Charter (1980), and in UN documents about basic human rights. Naess was doing something more subtle than many thought. He was not putting forth a single worldview and philosophy of life that everyone should adhere to in support of the international ecology movement. Instead, he was making an empirical claim based on overwhelming evidence that global social movements, from the grass roots up, consist of people with very diverse religious, philosophical, cultural, and personal orientations. Nonetheless, they can agree on certain courses of action and certain broad principles, especially at the international level. As supporters of a given movement, they can treat one another with mutual respect. Because of these misunderstandings Naess introduced an Apron Diagram to clearly illustrate his subtle distinctions. The Apron Diagram is meant to illustrate logical, as distinct from genetic, relations between views and their connection with social movements, policies and practical actions. By "logical relations" this means verbally articulated relations between the premises and conclusions. There is collective cooperation on global concerns, and yet a great variety of ultimate premises from which each person or group acts locally. Within global movements there is diversity at the local level because each place and community is different and must adapt to its unique setting. Thus, Naess stressed that his Ecosophy T is not meant to hold for everyone, since it is tailored to his very modest lifestyle suitable to a place such as Tvergastein. The ultimate premises for his whole view might be conceptually incompatible with those in someone else's whole views. But even if this is true, they could both support the Platform Principles of the deep ecology movement and other social-political global movements, such as for peace and social justice. In recognizing the principle that all living beings have intrinsic worth, one acknowledges they are good for their own sake. This does not commit one to biocentric equality or egalitarianism between species. Within the vast diversity of living beings, there are complex relationships the range of which is predation, competition, cooperation, and symbiosis. Many think that symbiosis and complementarity are important values to embrace as they are consistent with global cooperation, community life, and support for the deep ecology movement Platform.

Ecofeminist Response

Ecofeminism and deep ecology have been in dialogue for some time now, and while the debate between them has been very fruitful over the years, the exploration of their relationship remains important. As valuable as our individually reconnecting with the natural world may be, and as much as such experiences should be encouraged, some have questioned whether this approach is sufficient, given the magnitude of the threat that human encroachment poses to the nonhuman world. With this in mind, the call has been made for a broader challenge to the dominant culture than deep ecological experience may offer. This call has come most forcefully from another school of environmental ethics: ecofeminism. While sharing with deep ecologists a general concern for biocentrism and an appreciation for personal interaction with nonhuman reality, ecofeminists have also offered some harsh criticism of Leopold, Callicott, and advocates of the deep ecological approach, as well as extensionists like Singer and Regan. Like deep ecology, ecofeminism is not a singular theory, and covers a broad range of thought; in rough terms, its criticism of other forms of nature ethics is grounded in an attempt to synthesize the insights of environmental ethics and animal advocacy with a feminist analysis of Western ethics and culture. The resulting attempt to rethink our relationship to the animal and nature casts speciesism and anthropocentrism as symptoms of a deeper patriarchy in the Western tradition that needs to be deconstructed before a successful animal ethic can be produced. As Josephine Donovan details, there is evidence to suggest a strong emotional and philosophical affinity among Anglo-American antivivisectionists and suffragettes, who viewed their causes as common responses to Enlightenment rationalism and scientism, and jointly sought a "feminizing" of culture that would liberate human and animal alike. Ecological feminism began as a critique and rejection of the western cultural worldview with its overemphasis on rationality, and linearity. It argued against a Cartesian science which elevated the material and objective above the spiritual and the subjective as appropriate ways of knowing the world. Like deep ecology, ecological feminism stresses the importance of experience, and personal experience at that. However, the ecofeminists seem to be talking about experience in a sense more related to bioregionalism than deep ecology.

Both ecofeminism and deep ecology put forward a new conceptualization of the self. Some ecofeminists, such as Marti Kheel, argue that self-realization and identification with all nature places too much emphasis on the whole, at the expense of the independent being. Similarly, some ecofeminists place more emphasis on the problem of androcentrism rather than anthropocentrism. To others, like Karen J. Warren, the domination of women is tethered conceptually and historically to the domination of nature. Ecofeminism denies abstract individualism and embraces the interconnectedness of the living world; relationships, including our relationship with non-human nature, are not extrinsic to our identity and are essential in defining what it means to be human. Warren argues that hierarchical classifications in general, such as racism or speciesism, are all forms of discrimination and are no different from sexism. Thus, anthropocentrism is simply another form of discrimination as a result of our flawed value structure and should be abolished.

Experiential Deep Ecologist Joanna Macy has attempted to avoid these conflicts and criticisms through her Work that Reconnects. By focussing Deep Ecology on the experience of the consciousness of personal depth within the participant, she speaks of "The Greening of the Self", which is part of the epochal journey of our times from an Egoic or Egotistical Self to an Ecological Self.

Links with Other Philosophies

Parallels have been drawn between deep ecology and other philosophies, in particular those of the animal rights movement, Earth First!, Deep Green Resistance, and anarcho-primitivism.

Peter Singer's 1975 book *Animal Liberation* critiqued anthropocentrism and put the case for animals to be given moral consideration. This can be seen as a part of a process of expanding the prevailing system of ethics to wider groupings. However, Singer has disagreed with deep ecology's belief in the intrinsic value of nature separate from questions of suffering, taking a more utilitarian stance. The feminist and civil rights movements also brought about expansion of the ethical system for their particular domains. Likewise deep ecology brought the whole of nature under moral consideration. The links with animal rights are perhaps the strongest, as "proponents of such ideas argue that 'All life has intrinsic value'".

Many in the radical environmental direct-action movement Earth First! claim to follow deep ecology, as indicated by one of their slogans *No compromise in defence of mother earth*. In particular, David Foreman, the co-founder of the movement, has also been a strong advocate for deep ecology, and engaged in a public debate with Murray Bookchin on the subject. Judi Bari was another prominent Earth Firster who espoused deep ecology. Many Earth First! actions have a distinct deep ecological theme; often these actions will be to save an area of old growth forest, the habitat of a snail or an owl, even individual trees. Actions are often symbolic or have other political aims. At one point Arne Næss also engaged in environmental direct action, though not under the Earth First! banner, when he chained himself to rocks in front of Mardalsfossen, a waterfall in a Norwegian fjord, in a successful protest against the building of a dam.

There are also anarchist currents in the movement, especially in the United Kingdom. For example, Robert Hart, pioneer of forest gardening in temperate climates, wrote the essay "Can Life Survive?" in *Deep Ecology & Anarchism*.

Robert Greenway and Theodore Roszak have employed the deep ecology platform as a means to argue for ecopsychology. Although ecopsychology is a highly differentiated umbrella that encompasses many practices and perspectives, its ethos is generally consistent with deep ecology. As this now almost forty-year-old "field" expands and continues to be reinterpreted by a variety of practitioners, social and natural scientists, and humanists, "ecopsychology" may change to include these novel perspectives.

Heidegger's critique of technology has certainly inspired environmentalist and postmodernist of our time. Deep ecologists, like Heidegger, allege that certain metaphysical presuppositions are responsible for ecological destruction, and also contend that any transformation can be brought about only through a renewed awareness about the world. Then the key to environmental crisis, require an ontological shift: from an anthropocentric and utilitarian understanding of world to an understanding which lets things be. A non-anthropocentric humanity would probably initiate attitudes, practices, and institutions that would exhibit respect and care for all beings.

References

- Kennet M., and Heinemann V, (2006) Green Economics, Setting the Scene. in International Journal of Green Economics, Vol 1 issue 1/2 (2006) Inderscience, Geneva
- Smith, Richard (2015). Green Capitalism: the god that failed. World Economics Association. pp. 55–61. ISBN 978-1-911156-22-2
- ""Green" Capitalism builds Big Solar: Should we throw the baby out with the bathwater?". CNS web. 2016-04-10. Retrieved 2017-02-26
- Costanza R; et al. (1998). "The value of the world's ecosystem services and natural capital1". Ecological Economics. 25 (1): 3–15. doi:10.1016/S0921-8009(98)00020-2
- Reinhardt, F. (1999) 'Market failure and the environmental policies of firms: economic rationales for 'beyond compliance' behavior.' Journal of Industrial Ecology 3(1), 9-21
- Griffiths, M.R; Lucas, J.R. (November 16, 2016). Value economics: the ethical implications of value for new economic thinking. Palgrave Macmillan UK. ISBN 9781137541864
- "Spash, C. L. (1999) The development of environmental thinking in economics. Environmental Values 8(4): 413-435." (PDF). Retrieved 2012-12-23
- Hardin, Garrett (1968-12-13). "The Tragedy of the Commons". Science. 162 (3859): 1243–1248. ISSN 0036-8075. PMID 17756331. doi:10.1126/science.162.3859.1243
- Prothero, Andrea; Fitchett, James (June 2000). "Greening Capitalism: Opportunities for a Green Commodity". Journal of Macromarketing. 20 (1): 46–55
- Georgescu-Roegen, Nicholas (1971). The Entropy Law and the Economic Process. (Full book accessible in hree parts at SlideShare). Cambridge, Massachusetts: Harvard University Press. ISBN 0674257804
- Brouwer, Roy (January 2000). "Environmental value transfer: state of the art and future prospects". Ecological Economics. 32 (1). doi:10.1016/S0921-8009(99)00070-1. Retrieved 11 November 2016
- Bailey, Ian (2007). "Neoliberalism, climate governance and the scalar politics of EU emissions trading". Royal Geographical Society. 39 (4): 431–442. doi:10.1111/j.1475-4762.2007.00770.x

- Mattson L. (1975). Book Review: Positional Analysis for Decision-Making and Planning by Peter Soderbaum. The Swedish Journal of Economics
- Knapp G, Roheim CA and Anderson JL (2007) The Great Salmon Run: Competition Between Wild And Farmed Salmon World Wildlife Fund. ISBN 0-89164-175-0
- Farber, Stephen; Constanza, Robert; Wilson, Matthew (June 2002). "Economic and ecological concepts for valuing ecosystem services". Ecological Economics. 41 (3). doi:10.1016/S0921-8009(02)00088-5. Retrieved 11 November 2016
- Prudham, Scott (2009). "Pimping climate change: Richard Branson, global warming, and the performance of green capitalism". Environment and Planning. 41: 1594–1613. doi:10.1068/a4071
- "Spash, C.L. (2011) Social ecological economics: Understanding the past to see the future. American Journal of Economics and Sociology 70, 340-375" (PDF). Retrieved 2014-01-07
- Voinov, Alexey (2008). Systems science and modeling for ecological economics (1st ed.). Amsterdam: Elsevier Academic Press. ISBN 0080886175
- "Spash, C. L. (2007) The economics of climate change impacts à la Stern: Novel and nuanced or rhetorically restricted? Ecological Economics 63(4): 706-713" (PDF). Archived from the original (PDF) on 2014-02-02. Retrieved 2012-12-23

7

Allied Aspects of Green Politics

The allied fields of green politics discussed in this section are food politics, climate movement, animal rights movement and environmental movement. The environmental movement is a diverse movement which centers on human rights, ecology and health. The chapter serves as a source to understand the allied fields of green politics.

Environmental Movement

Apollo 8's *Earthrise*, 24 December 1968

The environmental movement (sometimes referred to as the ecology movement), also including conservation and green politics, is a diverse scientific, social, and political movement for addressing environmental issues. Environmentalists advocate the sustainable management of resources and stewardship of the environment through changes in public policy and individual behavior. In its recognition of humanity as a participant in (not enemy of) ecosystems, the movement is centered on ecology, health, and human rights.

The environmental movement is an international movement, represented by a range of organizations, from the large to grassroots and varies from country to country. Due to its large membership, varying and strong beliefs, and occasionally speculative nature, the environmental movement is not always united in its goals. The movement also encompasses some other movements with a more specific focus, such as the climate movement. At its broadest, the movement includes private citizens, professionals, religious devotees, politicians, scientists, nonprofit organizations and individual advocates.

History

Early Awareness

Levels of air pollution rose during the Industrial Revolution, sparking the first modern environmental laws to be passed in the mid-19th century

Early interest in the environment was a feature of the Romantic movement in the early 19th century. The poet William Wordsworth had travelled extensively in the Lake District and wrote that it is a "sort of national property in which every man has a right and interest who has an eye to perceive and a heart to enjoy".

The origins of the environmental movement lay in the response to increasing levels of smoke pollution in the atmosphere during the Industrial Revolution. The emergence of great factories and the concomitant immense growth in coal consumption gave rise to an unprecedented level of air pollution in industrial centers; after 1900 the large volume of industrial chemical discharges added to the growing load of untreated human waste. Under increasing political pressure from the urban middle-class, the first large-scale, modern environmental laws came in the form of Britain's Alkali Acts, passed in 1863, to regulate the deleterious air pollution (gaseous hydrochloric acid) given off by the Leblanc process, used to produce soda ash.

Conservation Movement

Students from the forestry school at Oxford, on a visit to the forests of Saxony in the year 1892

The modern conservation movement was first manifested in the forests of India, with the practical application of scientific conservation principles. The conservation ethic that began to evolve included three core principles: that the human activity damaged the environment, that there was a civic duty to maintain the environment for future generations, and that scientific, empirically based methods should be applied to ensure this duty was carried out. Sir James Ranald Martin was prominent in promoting this ideology, publishing many medico-topographical reports that demonstrated the scale of damage wrought through large-scale deforestation and desiccation, and lobbying extensively for the institutionalization of forest conservation activities in British India through the establishment of Forest Departments. The Madras Board of Revenue started local conservation efforts in 1842, headed by Alexander Gibson, a professional botanist who systematically adopted a forest conservation program based on scientific principles. This was the first case of state management of forests in the world. Eventually, the government under Governor-General Lord Dalhousie introduced the first permanent and large-scale forest conservation program in the world in 1855, a model that soon spread to other colonies, as well the United States. In 1860, the Department banned the use shifting cultivation. Dr. Hugh Cleghorn's 1861 manual, *The forests and gardens of South India*, became the definitive work on the subject and was widely used by forest assistants in the subcontinent.

Sir Dietrich Brandis joined the British service in 1856 as superintendent of the teak forests of Pegu division in eastern Burma. During that time Burma's teak forests were controlled by militant Karen tribals. He introduced the "taungya" system, in which Karen villagers provided labour for clearing, planting and weeding teak plantations. He formulated new forest legislation and helped establish research and training institutions. The Imperial Forest School at Dehradun was founded by him.

Formation of Environmental Protection Societies

The late 19th century saw the formation of the first wildlife conservation societies. The zoologist Alfred Newton published a series of investigations into the *Desirability of establishing a 'Close-time' for the preservation of indigenous animals* between 1872 and 1903. His advocacy for legislation to protect animals from hunting during the mating season led to the formation of the Plumage League (later the Royal Society for the Protection of Birds) in 1889. The society acted as a protest group campaigning against the use of great crested grebe and kittiwake skins and feathers in fur clothing. The Society attracted growing support from the suburban middle-classes, and influenced the passage of the Sea Birds Preservation Act in 1869 as the first nature protection law in the world.

For most of the century from 1850 to 1950, however, the primary environmental cause was the mitigation of air pollution. The Coal Smoke Abatement Society was formed in 1898 making it one of the oldest environmental NGOs. It was founded by artist Sir William Blake Richmond, frustrated with the pall cast by coal smoke. Although there were earlier pieces of legislation, the Public Health Act 1875 required all furnaces and fireplaces to consume their own smoke.

Systematic and general efforts on behalf of the environment only began in the late 19th century; it grew out of the amenity movement in Britain in the 1870s, which was a reaction to industrialization, the growth of cities, and worsening air and water pollution. Starting with the formation of the Commons Preservation Society in 1865, the movement championed rural preservation against the encroachments of industrialisation. Robert Hunter, solicitor for the society, worked with Hardwicke Rawnsley, Octavia Hill, and John Ruskin to lead a successful campaign to prevent the construction of railways to carry slate from the quarries, which would have ruined the unspoilt valleys of Newlands and Ennerdale. This success led to the formation of the Lake District Defence Society (later to become The Friends of the Lake District).

John Ruskin an influential thinker who articulated the Romantic ideal of environmental protection and conservation

In 1893 Hill, Hunter and Rawnsley agreed to set up a national body to coordinate environmental conservation efforts across the country; the "National Trust for Places of Historic Interest or Natural Beauty" was formally inaugurated in 1894. The organisation obtained secure footing through the 1907 National Trust Bill, which gave the trust the status of a statutory corporation. and the bill was passed in August 1907.

An early "Back-to-Nature" movement, which anticipated the romantic ideal of modern environmentalism, was advocated by intellectuals such as John Ruskin, William Morris, and Edward Carpenter, who were all against consumerism, pollution and other activities that were harmful to the natural world. The movement was a reaction to the urban conditions of the industrial towns, where sanitation was awful, pollution levels intolerable and housing terribly cramped. Idealists championed the rural life as a mythical Utopia and advocated a return to it. John Ruskin argued that people should return to a *small piece of English ground, beautiful, peaceful, and fruitful. We will have no steam engines upon it . . . we will have plenty of flowers and vegetables . . . we will have some music and poetry; the children will learn to dance to it and sing it.*

Practical ventures in the establishment of small cooperative farms were even attempted and old rural traditions, without the "taint of manufacture or the canker of artificiality", were enthusiastically revived, including the Morris dance and the maypole.

The movement in the United States began in the late 19th century, out of concerns for protecting the natural resources of the West, with individuals such as John Muir and Henry David Thoreau making key philosophical contributions. Thoreau was interested in peoples' relationship with nature and studied this by living close to nature in a simple life. He published his experiences in the book *Walden*, which argues that people should become intimately close with nature. Muir came to believe in nature's inherent right, especially after spending time hiking in Yosemite Valley and studying both the ecology and geology. He successfully lobbied congress to form Yosemite National Park and went on to set up the Sierra Club in 1892. The conservationist principles as well as the belief in an inherent right of nature were to become the bedrock of modern environmentalism. However, the early movement in the U.S. developed with a contradiction; preservationists like John Muir wanted land and nature set aside for its own sake, and conservationists, such as Gifford Pinchot (appointed as the first Chief of the US Forest Service from 1905-1910), wanted to manage natural resources for human use.

WALDEN;

OR,

LIFE IN THE WOODS.

BY HENRY D. THOREAU,

AUTHOR OF "A WEEK ON THE CONCORD AND MERRIMACK RIVERS."

I do not propose to write an ode to dejection, but to brag as lustily as chanticleer in the morning, standing on his roost, if only to wake my neighbors up. — Page 92.

BOSTON:

TICKNOR AND FIELDS.

M DCCC LIV.

Original title page of *Walden* by Henry David Thoreau

20th Century

In the 20th century, environmental ideas continued to grow in popularity and recognition. Efforts were beginning to be made to save the wildlife, particularly the American bison. The death of the last passenger pigeon as well as the endangerment of the American bison helped to focus the minds of conservationists and popularize their concerns. In 1916 the National Park Service was founded by US President Woodrow Wilson. Pioneers of the movement called for more efficient and professional management of natural resources. They fought for reform because they believed the destruction of forests, fertile soil, minerals, wildlife and water resources would lead to the downfall of society. The group that has been the most active in recent years is the climate movement.

The conservation of natural resources is the fundamental problem. Unless we solve that problem, it will avail us little to solve all others.

Theodore Roosevelt (4 October 1907)

The U.S movement began to take off after World War II as people began to recognize the costs of environmental negligence, disease, and the expansion of air and water pollution through the occurrence of several environmental disasters that occurred post-World War II. Aldo Leopold wrote "A Sand County Almanac" in the 1940s. He believed in a land ethic that recognized that maintaining the "beauty, integrity, and health of natural systems" as a moral and ethical imperative.

Another major literary force in the promotion of the environmental movement was Rachel Carson's "Silent Spring" about declining bird populations due to DDT, an insecticide, pollution and man's attempts to control nature through use of synthetic substances. Her core message for her readers, was to identify the complex and fragile ecosystem and the threats facing the people. In 1958 Carson started to work on her last book, with an idea that nature needs human protection. Her influence was radioactive fallout, smog, food additives, and pesticide use. Carson's main focus was on pesticides, which led her to identify nature ad fragile and the use of technology dangerous to humans and other species.

Both of these books helped bring the issues into the public eye Rachel Carson's "Silent Spring" sold over two million copies.

Earth Day 2007 at City College, San Diego

The first Earth Day was celebrated on 22 April 1970. Its founder, former Wisconsin Senator, Gaylord Nelson was inspired to create this day of environmental education and awareness after seeing the oil spill off the coast of Santa Barbara in 1969. Greenpeace was created in 1971 as an organization that believed that political advocacy and legislation were ineffective or inefficient solutions and supported non-violent action. 1980 saw the creation of Earth First!, a group with an ecocentric view of the world – believing in equality between the rights of humans to flourish, the rights of all other species to flourish and the rights of life-sustaining systems to flourish.

In the 1950s, 1960s, and 1970s, several events illustrated the magnitude of environmental damage caused by humans. In 1954, a hydrogen bomb test at Bikini Atoll exposed the 23 man crew of the Japanese fishing vessel *Lucky Dragon 5* to radioactive fallout. In 1967 the oil tanker *Torrey Canyon* ran aground off the coast of Cornwall, and in 1969 oil spilled from an offshore well in California's Santa Barbara Channel. In 1971, the conclusion of a lawsuit in Japan drew international attention to the effects of decades of mercury poisoning on the people of Minamata.

At the same time, emerging scientific research drew new attention to existing and hypothetical threats to the environment and humanity. Among them were Paul R. Ehrlich, whose book *The Population Bomb* (1968) revived Malthusian concerns about the impact of exponential population growth. Biologist Barry Commoner generated a debate about growth, affluence and "flawed technology." Additionally, an association of scientists and political leaders known as the Club of Rome published their report *The Limits to Growth* in 1972, and drew attention to the growing pressure on natural resources from human activities.

Meanwhile, technological accomplishments such as nuclear proliferation and photos of the Earth from outer space provided both new insights and new reasons for concern over Earth's seemingly small and unique place in the universe.

In 1972, the United Nations Conference on the Human Environment was held in Stockholm, and for the first time united the representatives of multiple governments in discussion relating to the state of the global environment. This conference led directly to the creation of government environmental agencies and the UN Environment Program.

By the mid-1970s anti-nuclear activism had moved beyond local protests and politics to gain a wider appeal and influence. Although it lacked a single co-ordinating organization the anti-nuclear movement's efforts gained a great deal of attention, especially in the United Kingdom and United States. In the aftermath of the Three Mile Island accident in 1979, many mass demonstrations took place. The largest one was held in New York City in September 1979 and involved 200,000 people.

Since the 1970s, public awareness, environmental sciences, ecology, and technology have advanced to include modern focus points like ozone depletion, global climate change, acid rain, and the potential of genetically modified organisms (GMOs). Some argue that genetically modified plants and animals are inherently bad because they are unnatural. Others point out the possible benefits of GM crops such as water conservation through corn modified to be less "thirsty" and decreased pesticide use through insect - resistant crops.

United States

Beginning in the conservation movement at the beginning of the 20th century, the contemporary environmental movement's roots can be traced back to Murray Bookchin's *Our Synthetic Environment*, Paul R. Ehrlich's The Population Bomb, and Rachel Carson's *Silent Spring*. American environmentalists have campaigned against nuclear weapons and nuclear power in 1960s and 1970s, acid rain in the 1980s, ozone depletion and deforestation in the 1990s, and most recently climate change and global warming.

The United States passed many pieces of environmental legislation in the 1970s, such as the Clean Water Act, the Clean Air Act, the Endangered Species Act, and the National Environmental Policy Act. These remain as the foundations for current environmental standards.

US Environmental History

- 1832- Hot Springs Reservation
- 1864- Yosemite Valley
- 1872- Yellowstone National Park
- 1892- Sierra Club
- 1916- National Park Service Organic Act
- 1916- National Audubon Society
- 1949- UN Scientific Conference on the Conservation and Utilization of Resources
- 1961- World Wildlife Foundation
- 1964- Land and Water Conservation Act
- 1964- National Wilderness Preservation System

- 1968- National Trails System Act
- 1968- National Wild and Scenic Rivers System/Wild and Scenic Rivers Act
- 1969- National Environmental Policy Act
- 1970- First Earth Day- 22 April
- 1970- Clean Air Act
- 1970- Environmental Protection Agency
- 1971- Greenpeace
- 1972- Clean Water Act
- 1973- Endangered Species Act
- 1980- Earth First!
- 1992- UN Earth Summit in Rio de Janeiro

Latin America

After the International Environmental Conference in Stockholm in 1972 Latin American officials returned with a high hope of growth and protection of the fairly untouched natural resources. Governments spent millions of dollars, and created departments and pollution standards. However, the outcomes have not always been what officials had initially hoped. Activists blame this on growing urban populations and industrial growth. Many Latin American countries have had a large inflow of immigrants that are living in substandard housing. Enforcement of the pollution standards is lax and penalties are minimal; in Venezuela, the largest penalty for violating an environmental law is 50,000 bolivar fine ($3,400) and 3 days in jail. In the 1970s or 1980s many Latin American countries were transitioning from military dictatorships to democratic governments.

Brazil

In 1992, Brazil came under scrutiny with the United Nations Conference on Environment and Development in Rio de Janeiro. Brazil has a history of little environmental awareness. It has the highest biodiversity in the world and also the highest amount of habitat destruction. One-third of the world's forests lie in Brazil, and they have the largest river, The Amazon, and the largest rainforest, the Amazon Rainforest. The people have raised funds to create state parks and increase the consciousness of people who have destroyed forests and polluted waterways. They have several organizations that have fronted the environmental movement. The Blue Wave Foundation was created in 1989 and has partnered with advertising companies to promote national education campaigns to keep Brazil's beaches clean. Funatura was created in 1986 and is a wildlife sanctuary program. Pro-Natura International is a private environmental organization created in 1986.

Europe

In 1952 the Great London Smog episode killed thousands of people and led the UK to create the first Clean Air Act in 1956. In 1957 the first major nuclear accident occurred in Windscale in northern England. The supertanker *Torrey Canyon* ran aground off the coast of Cornwall in 1967 causing the first major oil leak that killed marine life along the coast. In 1972, in Stockholm, the United Nations Conference on the Human Environment created the UN Environment Programme. The EU's environmental policy was formally founded by a European Council declaration and the first five-year environment programme was adopted. The main idea of the declaration was that prevention is better than the cure and the polluter should pay. 1979 saw the partial meltdown of Three Mile Island in the USA.

In the 1980s the green parties that were created a decade before began to have some political success.. In 1986, there was a nuclear accident in Chernobyl, Ukraine. The end of the 1980s and start of the 1990s saw the fall of communism across central and Eastern Europe, the fall of the [Berlin Wall], and the Union of East and West Germany. In 1992 there was a UN summit held in Rio de Janeiro where Agenda 21 was adopted. The Kyoto Protocol was created in 1997 which set specific targets and deadlines to reduce global greenhouse gas emissions. In the early 2000s activists believed that environmental policy concerns were overshadowed by energy security, globalism, and terrorism.

Asia

Middle East

The environmental movement is reaching the less developed world with different degrees of success. The Arab world, including the Middle East and North Africa, has different adaptations of the environmental movement. Countries on the Persian Gulf have high incomes and rely heavily on the large amount of energy resources in the area. Each country in the Arab world has varying combinations of low or high amounts of natural resources and low or high amounts of labor.

The League of Arab States has one specialized sub-committee, of 12 standing specialized subcommittee in the Foreign Affairs Ministerial Committees, which deals with Environmental Issues. Countries in the League of Arab States have demonstrated an interest in environmental issue, on paper some environmental activists have doubts about the level of commitment to environmental issues;; being a part of the world community may have obliged these countries to portray concern for the environment. Initial level of environmental awareness may be the creation of a ministry of the environment. The year of establishment of a ministry is also indicative of level of engagement. Saudi Arabia was the first to establish environmental law in 1992 followed by Egypt in 1994. Somalia is the only country without environmental law. In 2010 the Environmental Performance Index listed Algeria as the top Arab country at 42 of 163; Morocco was at 52 and Syria at 56. The Environmental Performance Index measures the ability of a country to actively manage and protect their environment and the health of their citizens. A weighted index is created by giving 50% weight for environmental health objective (health) and 50% for ecosystem vitality (ecosystem); values range from 0-100. No Arab countries were in the top quartile, and 7 countries were in the lowest quartile.

South Korea and Taiwan

South Korea and Taiwan experienced similar growth in industrialization from 1965-1990 with few environmental controls. South Korea's Han River and Nakdong River were so polluted by unchecked dumping of industrial waste that they were close to being classified as biologically dead. Taiwan's formula for balanced growth was to prevent industrial concentration and encourage manufacturers to set up in the countryside. This led to 20% of the farmland being polluted by industrial waste and 30% of the rice grown on the island was contaminated with heavy metals. Both countries had spontaneous environmental movements drawing participants from different classes. Their demands were linked with issues of employment, occupational health, and agricultural crisis. They were also quite militant; the people learned that protesting can bring results. The polluting factories were forced to make immediate improvements of the conditions or pay compensation to victims. Some were even forced to shut down or move locations. The people were able to force the government to come out with new restrictive rules on toxins, industrial waste, and air pollution. All of these new regulations caused the migration of those polluting industries from Taiwan and South Korea to China and other countries in Southeast Asia with more relaxed environmental laws.

China

China's environmental movement is characterized by spontaneous alliances that often only occur at the local level. The Chinese have realized the ability of riots and protests to have success and had led to an increase in disputes in China by 30% since 2005 to more than 50,000 events. Protests cover topics such as environmental issues, land loss, income, and political issues. They have also grown in size from about 10 people or fewer in the mid-1990s to 52 people per incident in 2004. China has more relaxed environmental laws than other countries in Asia, so many polluting factories have relocated to China causing pollution in China. Water pollution, water scarcity, soil pollution, soil degradation, and desertification are issues currently in discussion in China. The groundwater table of the North China Plain is dropping by 1.5 m (5 ft) per year. This groundwater table occurs in the region of China that produces 40% of the country's grain.

India

Environmental and public health is an ongoing struggle within India. The first seed of an environmental movement in India was the foundation in 1964 of *Dasholi Gram Swarajya Sangh*, a labour cooperative started by Chandi Prasad Bhatt. It was inaugurated by Sucheta Kriplani and founded on a land donated by Shyma Devi. This initiative was eventually followed up with the Chipko movement starting in 1974.

The most severe single event underpinning the movement was the Bhopal gas leakage on 3 December 1984. 40 tons of methyl isocyanate was released, immediately killing 2,259 people and ultimately affecting 700,000 citizens.

India has a national campaign against Coca-Cola and Pepsi Cola plants due to their practices of drawing ground water and contaminating fields with sludge. The movement is characterized by local struggles against intensive aquaculture farms. The most influential part of the environmental movement in India is the anti-dam movement. Dam creation has been thought of as a way for India to catch up with the West by connecting to the power grid with giant dams, coal or oil-powered plants, or nuclear plants. Jhola Aandolan a mass movement is conducting as fighting against polyethylene carry bags uses and promoting cloth/jute/paper carry bags to protect environment & nature. Activists in the Indian environmental movement consider global warming, sea levels rising, and glaciers retreating decreasing the amount of water flowing into streams to be the biggest challenges for them to face in the early twenty-first century. Eco Revolution movement has been started by Eco Needs Foundation in 2008 from Aurangabad Maharashtra state .The pioneer of Eco

Revolution movement is Priyanand Agale more than 20,000 youths associated with this movement to strengthen this movement Eco Needs Foundation organized mass environmental awareness programmes. To sought participation of children,Youths, researchers, spiritual and political leaders.Foundation had conducted International conferences at India Eco Revolution 2011 which concluded with Aurangabad Declaration for River Conservation. Eco Revolution 2012 conference was conducted in Sri Lanka which concluded with Colombo Declaration on spirituality for environmental conservation in collaboration with Sri Lanka government. Eco Revolution 2013 was conducted at Nepal which concluded with the Phokhara Declaration for effect of climate change at high altitude .Foundation launched the worlds first environmental social networking site ecoface. in. Foundation developed model of sustainable development at Dhanora village of Dholpur, Rajasthan as a India's First smart village.

Bangladesh

Mithun Roy Chowdhury, President, Save Nature & Wildlife (SNW), Bangladesh, insisted that the people of Bangladesh raise their voice against Tipaimukh Dam, being constructed by the Government of India. He said Tipaimukh Dam project will be another "death trap for Bangladesh like the Farakka Barrage," that would lead to an environmental disaster for 50 million people in the Meghna River basin. He said that this project will start desertification in Bangladesh.

Scope of the Movement

Before flue-gas desulfurization was installed, the air-polluting emissions from this power plant in New Mexico contained excessive amounts of sulfur dioxide

Environmental science is the study of the interactions among the physical, chemical and biological components of the environment.

- Ecology, or ecological science, is the scientific study of the distribution and abundance of living organisms and how these properties are affected by interactions between the organisms and their environment.

Primary Focus Points

The environmental movement is broad in scope and can include any topic related to the environment, conservation, and biology, as well as preservation of landscapes, flora, and fauna for a variety of purposes and uses. See List of environmental issues. When an act of violence is committed against someone or some institution in the name of environmental defense it is referred to as eco terrorism.

- The conservation movement seeks to protect natural areas for sustainable consumption, as well as traditional (hunting, fishing, trapping) and spiritual use.
- Environmental conservation is the process in which one is involved in conserving the natural aspects of the environment. Whether through reforestation, recycling, or pollution control, environmental conservation sustains the natural quality of life.
- Environmental health movement dates at least to Progressive Era, and focuses on urban standards like clean water, efficient sewage handling, and stable population growth. Environmental health could also deal with nutrition, preventive medicine, aging, and other concerns specific to human well-being. Environmental health is also seen as an indicator for the state of the environment, or an early warning system for what may happen to humans
- Environmental justice is a movement that began in the U.S. in the 1980s and seeks an end to environmental racism and prevent low-income and minority communities from an unbalanced exposure to highways, garbage dumps, and factories. The Environmental Justice movement seeks to link "social" and "ecological" environmental concerns, while at the same time preventing de facto racism, and classism. This makes it particularly adequate for the construction of labor-environmental alliances.
- Ecology movement could involve the Gaia Theory, as well as Value of Earth and other interactions between humans, science, and responsibility.
- Bright green environmentalism is a currently popular sub-movement, which emphasizes the idea that through technology, good design and more thoughtful use of energy and resources, people can live responsible, sustainable lives while enjoying prosperity.
- Light green, and dark green environmentalism are yet other sub-movements, respectively distinguished by seeing environmentalism as a lifestyle choice (light greens), and promoting reduction in human numbers and/or a relinquishment of technology (dark greens)
- Deep Ecology is an ideological spinoff of the ecology movement that views the diversity and integrity of the planetary ecosystem, in and for itself, as its primary value.
- The anti-nuclear movement opposes the use of various nuclear technologies. The initial anti-nuclear objective was nuclear disarmament and later the focus began to shift to other issues, mainly opposition to the use of nuclear power. There have been many large anti-nuclear demonstrations and protests. Major anti-nuclear groups include Campaign for Nuclear Disarmament, Friends of the Earth, Greenpeace, International Physicians for the Prevention of Nuclear War, and the Nuclear Information and Resource Service.

Environmental Law and Theory

Property Rights

Many environmental lawsuits question the legal rights of property owners, and whether the general public has a right to intervene with detrimental practices occurring on someone else's land. Environmental law organizations exist all across the world, such as the Environmental Law and Policy Center in the midwestern United States.

Citizens' Rights

One of the earliest lawsuits to establish that citizens may sue for environmental and aesthetic harms was Scenic Hudson Preservation Conference v. Federal Power Commission, decided in 1965 by the Second Circuit Court of Appeals. The case helped halt the construction of a power plant on Storm King Mountain in New York State.

Nature's Rights

Christopher D. Stone's 1972 essay, "Should trees have standing?" addressed the question of whether natural objects themselves should have legal rights. In the essay, Stone suggests that his argument is valid because many current rightsholders (women, children) were once seen as objects.

Environmental Reactivism

Numerous criticisms and ethical ambiguities have led to growing concerns about technology, including the use of potentially harmful pesticides, water additives like fluoride, and the extremely dangerous ethanol-processing plants.

NIMBY syndrome refers to public outcry caused by knee-jerk reaction to an unwillingness to be exposed to even necessary developments. Some serious biologists and ecologists created the scientific ecology movement which would not confuse empirical data with visions of a desirable future world.

Environmentalism Today

Composite images of Earth generated by NASA in 2001 (left) and 2002 (right)

Today, the sciences of ecology and environmental science, in addition to any aesthetic goals, provide the basis of unity to some of the serious environmentalists. As more information is gathered in scientific fields, more scientific issues like biodiversity, as opposed to mere aesthetics, are a concern to environmentalists. Conservation biology is a rapidly developing field.

In recent years, the environmental movement has increasingly focused on global warming as one of the top issues. As concerns about climate change moved more into the mainstream, from the connections drawn between global warming and Hurricane Katrina to Al Gore's film An Inconvenient Truth, more and more environmental groups refocused their efforts. In the United States, 2007 witnessed the largest grassroots environmental demonstration in years, Step It Up 2007, with rallies in over 1,400 communities and all 50 states for real global warming solutions.

Many religious organizations and individual churches now have programs and activities dedicated to environmental issues. The religious movement is often supported by interpretation of scriptures. Most major religious groups are represented including Jewish, Islamic, Anglican, Orthodox, Evangelical, Zoroastrian, Christian and Catholic.

Radical Environmentalism

Radical environmentalism emerged from an ecocentrism-based frustration with the co-option of mainstream environmentalism. The radical environmental movement aspires to what scolar Christopher Manes calls "a new kind of environmental activism: iconoclastic, uncompromising, discontented with traditional conservation policy, at times illegal ..." Radical environmentalism presupposes a need to reconsider Western ideas of religion and philosophy (including capitalism, patriarchy and globalization) sometimes through "resacralising" and reconnecting with nature. Greenpeace represents an organisation with a radical approach, but has contributed in serious ways towards understanding of critical issues, and has a science-oriented core with radicalism as a means to media exposure. Groups like Earth First! take a much more radical posture. Some radical environmentalist groups, like Earth First! and the Earth Liberation Front, illegally sabotage or destroy infrastructural capital.

Criticisms

Conservative critics of the movement characterize it as radical and misguided. Especially critics of the United States Endangered Species Act, which has come under scrutiny lately, and the Clean Air Act, which they said conflict with private property rights, corporate profits and the nation's overall economic growth. Critics also challenge the scientific evidence for global warming. They argue that the environmental movement has diverted attention from more pressing issues.

Food Politics

Food politics are the political aspects of the production, control, regulation, inspection, distribution and consumption of food. The politics can be affected by the ethical, cultural, medical and environmental disputes concerning proper farming, agricultural and retailing methods and regulations.

Policy

Government policies around food production, distribution, and consumption influence the cost, availability, and safety of the food supply domestically and internationally. On a national scale, food policy work affects farmers, food processors, wholesalers, retailers and consumers. Commodity crops, such as corn, rice, wheat, and soy are most often at the heart of agricultural policy-making. While most food policy is initiated domestically, there are international ramifications. Globally, protectionist trade policies, international trade agreements, famine, political instability, and development aid are among the primary influences on food policy. Increasingly, climate change concerns and predictions are gaining the attention of those most concern with ensuring an adequate worldwide food supply.

Food Politics in the U.S.

A number of contemporary issues around food policy issues have surfaced in the United States due to changes in the production of food and concerns about the nutritional quality of commercially prepared foods.

Technology

As with many industries, the food industry has experienced growth in the capacity to produce food with the use of improved technologies. In developed countries, there are a number of important trends at play. Yields, or the amount of food harvested per acre of cropland, have increased less than one percent per year in since at least the 1960s and the amount of land devoted to crop use is in decline due to development pressures for housing and other economic concerns. In the U.S. alone, about 3000 acres of productive farmland are lost each day. This places a premium on quality yields from existing acres of farmland. In addition, the demand for meat products worldwide, expected to double by 2020, has accelerated a trend toward raising more animals on fewer acres of land.

Farming of Animals

More intensive forms of animal farming have largely replaced traditional methods of raising pigs, cattle, poultry and fish for human consumption in the U.S. The increased development of Concentrated Animal Feeding Operations have been associated with increased risk of foodborne illnesses from e.coli, environmental degradation, and increased emissions of ammonia, carbon dioxide, and methane into the air. In addition to food safety and environmental concerns, organizations such as the People for the Ethical Treatment of Animals and the Humane Society of the United States (HSUS) have drawn attention to a range of practices that allow for the more efficient raising of animals for meat consumption, but these practices stress the animals, the land on which they are raised and the supply of food for human consumption. In a recent report on industrialized animal agriculture, HSUS called on people in Western countries to shift to a plant-based diet because half the world's grain crop is used to raise animals for meat, eggs, and milk. Fish farming has also come under scrutiny due to high concentrations of fish in smaller spaces than is experienced in the wild. For both land and water animals, the prophylactic use of antibiotics to promote growth and stem the spread of infection among the animals has also been questioned due to concerns that this practice may contribute to strains of antibiotic-resistant bacteria.

Genetically Modified Foods

The use of genetically modified seeds to grow commodity and other crops in the U.S. has drawn criticism from organizations such as Greenpeace, The Non-GMO Project, and the Organic Consumers Association among others. Concerns center on both food safety and the erosion of agricultural biodiversity. While the European Union regulates genetically engineered foods as they would any other new product requiring extensive testing to provide it is safe for human consumption, the U.S. does not. The Food and Drug Administration generally considers a food with origins from genetically modified organisms (GMO) to be as safe as its conventional counterpart. Numerous studies have backed industry claims that GMO foods appear to be safe for human consumption, including an examination of more than 130 research projects conducted in the European Union prior to 2010 and work published by the American Medical Association's Council on Science and Public Health.

In the U.S., the political debate has centered primarily on whether or not to label products with GMO origins to better inform the public about the content of the foods they purchase. A statewide ballot question that would have mandated labeling of GMO products in California was defeated in 2012. The measure, known as Proposition 37 was leading by a wide margin in early polling but was defeated after an advertising blitz bankrolled by Monsanto, the largest supplier of genetically engineered seeds, Kraft Foods, Coca-Cola, PepsiCo., and other large food business interests. The ballot question's results were closely watched across the country as advocates for the measure hoped that it would pass and spur the federal government to mandate labeling of GMO foods as well. In the wake of the labeling law proposal's defeat, an organization called March Against Monsanto was formed to continue to keep alive the public debate about labeling GMO food products. In 2013, a ballot initiative that would have required labels on GMO foods sold in the state of Washington was defeated by voters, again after a campaign against the initiative was led by major food companies.

Pesticide Use

Among the much-heralded impacts of the Green Revolution was the spread of technological advances in the development of pesticides to ensure higher crop yields. Health effects of pesticides have led to a number of regulatory and non-regulatory efforts to control potential harm to human health from these chemicals in the food supply. The US Environmental Protection Agency has jurisdiction of the use of pesticides in crop management and sets tolerances for trace amounts of pesticides that may be found in the food supply. About 12,000 samples of fruits and vegetables available to U.S. consumers are collected each year and tested for residue from pesticides and the results are published in an annual Pesticide Data Program (PDP) hosted by the USDA. The non-profit organization, Environmental Working Group, publishes an annual list called the Dirty Dozen designed to direct consumers' attention to the fruits and vegetables most contaminated by pesticide residue. In its annual Shopper's Guide to Pesticides, the organization urges shoppers to continue eating fresh produce, but consider purchasing organically grown versions of the 12 most contaminated items.

"Big food"

Food manufacturing and processing is a heavily concentrated industry. The 10 largest food companies in the United States control more than half of all food sales domestically and a growing percentage of packaged food and beverage products on store shelves worldwide. Ranked by food sales, PepsiCo, Inc., is the largest food manufacturer in the U.S., followed by Tyson Foods, Nestlé, JBS USA, and Anheuser-Busch, according to a 2013 list published by *Food Processing* magazine. According to figures from the United States Census Bureau from 2007, the most highly concentrated food industries in the country included cane sugar refining, breakfast cereals, bottled water, and cookie/cracker manufacturing using the 4-firm concentration ratio. Consolidation of this industry took place in the 1970s and 1980s through a series of mergers and acquisitions.

"Big food" has come under fire not only because a small number of players are responsible for a large percentage of the food supply chain, but because of concerns about the links between the highly processed foods they produce and the obesity epidemic both in the U.S. and worldwide. The director general of the World Health Organization, in a speech given at the 8th Global Conference on Health Promotion in Helsinki, Finland in June 2013, noted that the public health community's efforts to combat chronic diseases such as diabetes, hypertension and cardiovascular disease are pitted against the economic interests of the powerful food industry. Several studies are exploring processed foods with high concentrations of sugar, refined carbohydrates, fat, salt and caffeine for addictive properties.

Marketing and other strategies of the food industry have been compared to those of the tobacco industry at the height of its influence in the consumer marketplace. In response, the food industry has engaged in some voluntary efforts to improve the nutritional content of their foods. In 2005, General Mills announced a plan to ensure that all of its breakfast cereals contained at least eight grams of whole grain per serving. In 2006, Campbell Soup Company announced an initiative to reduce sodium in its products by at least 25 percent. Due to slumping sales, Campbell's acknowledged that it was adding more sodium back into some of its soups in 2011.

Food Movements

A cultural backlash against an increasingly mechanized food industry has taken a number of different forms.

- Local food is a loosely defined term that describes a movement to shift food expenditures by individuals, families, community organizations, schools, restaurants, and other institutions from foods produced and shipped long distances by larger corporate entities to regional farmers and other local producers of food. Small farming interests, relatively heterogeneous products and short supply chains characterize local food markets, though there is no agreed-upon measure of the distances that constitute "local." Community-supported agriculture is a mechanism for connecting consumers with local farmers. Farm-to-table efforts are also part of the local food movement.
- Meatless Monday is a public health awareness campaign encouraging individuals and families to eat a meat-free diet at least once each week. Launched in 2003 through the Johns Hopkins Center for a Livable Future, its focus is on preventable diseases associated with excessive meat consumption but the campaign has also been incorporated by many concerned about sustainable agriculture and the environment.
- Slow Food is an international movement founded in Italy in 1986, with Slow Food USA established in 2000. The organization stands in opposition to "the standardization of taste and culture, and the unrestrained power of food industry multinationals and industrial agriculture."

Among those influential in the food movement in the United States are writers, including Michael

Pollan and Marion Nestle, and celebrity chefs such as Alice Waters, Mario Batali and Jamie Oliver. Popular books and movies on contemporary topics in food include Fast Food Nation, The Omnivore's Dilemma and the documentary Food, Inc.. In 2011, the president of the American Farm Bureau Federation referred to this influential group as "self-appointed food elitists" and the Washington Post published an op-ed from Eric Schlosser, author of Fast Food Nation, defending the work he and colleagues have done to improve food systems in the United States.

Social Justice

While the production and distribution of food is primarily an economic activity, advocates for a variety of social justice concerns are increasingly aware of the role that food policy plays in issues of greatest concern to causes they espouse.

Biofuel Mandates and Food Supply

The interests of varied sectors of the agricultural industry are not always in alignment as illustrated by tensions stemming from a drought in 2012 that affected domestic corn production. Described as the most severe and extensive drought in the U.S. in the last 25 years by the USDA Economic Research Service, thousands of acres of mostly corn and soy fields in the Midwest were damaged or destroyed. This led to increased pressure on the federal government from some domestic farmers and international anti-hunger organizations to relax the Renewable Fuel Standard that call for a portion of the US-grown corn supply to be set aside for ethanol production. Meat and poultry producers, both of whom rely on corn for animal feed and feared rising prices due to the drought conditions, accused the federal government of "picking winners and losers" with its ethanol policy while ethanol producers, many of whom are corn farmers, argued that price spikes would affect them as well and that ethanol production had been scaled back.

Domestic Food Aid

Offering government food assistance to the lowest income Americans dates back to the administration of President Franklin D. Roosevelt and has continued into the 21st century. In FY 2011, the budget for the USDA Food and Nutrition Service, which is responsible for the major feeding programs, was $107 billion. The largest single food assistance program in the country is the Supplemental Nutrition Assistance Program, the provisions for which are contained in a Farm Bill that is re-authorized by Congress and signed by the president every five years. Benefits to SNAP recipients cost approximately $75 billion in 2012. Largely uncontroversial for most of its history, the SNAP program was targeted for major cuts by members of the House of Representatives in the 2012 Farm Bill re-authorization attempt. House leaders also endeavored to separate the SNAP program from the Farm Bill, splitting the long-standing coalition of urban and rural legislators who traditionally backed the renewals of funding for the Farm Bill every five years.

Increases in the size of the SNAP caseload during the early 2000s were associated with increases in the unemployment rate and with a number of policy changes made to the program in many states. A series of six measures to better understand employment trends developed by the Bureau of Labor Statistics, three of which are more conservative estimates of unemployment and three of which define it more broadly, all showed correlations with SNAP participation. In particular, it was suggested that longer-term unemployment results in the heaviest utilization of SNAP benefits.

In addition to concerns about the cost of the program from fiscal conservatives, leaders in the movement to improve the nutritional content of the American diet suggested changes to the program to preclude the purchase of sugar-sweetened soft drinks or other forms of junk food with low nutritional value. In fact, the House of Representatives' version of the initial Food Stamp Act of 1964 prohibited using food stamps for the purchase of soft drinks, but the provision was not adopted. Efforts to more narrowly define the food purchases by SNAP recipients were derided by anti-hunger organizations as a form of paternalism. The 2008 Farm Bill re-authorization established the Healthy Incentives Pilot, a $20 million effort in five states to learn if offering select SNAP recipients credits on purchases of fresh, frozen, canned, or dried fruits and vegetables with no added sugar, salt, fat, or oil will result in increased purchases of these foods. Results of the pilot study were expected in 2014.

In addition to advocacy work in Washington, D.C., on behalf of those in poverty, public awareness campaigns around the constraints faced by families receiving SNAP were launched. The food stamp challenge or SNAP challenge is one mechanism used by advocates such as Feeding America. Individuals are challenged to restrict food spending for a week to levels typical for families receiving SNAP benefits.

Labor and Immigration

Hired farmworkers are among the most economically disadvantaged groups in the United States. Farm labor statistics are published twice yearly by the National Agricultural Statistics Service, a division of the U.S. Department of Agriculture and are compared with similar data from the prior year. In April 2013, the number of workers hired by farm operators was 732,000. Field workers received an average pay of $10.92 per hour and livestock workers earned $11.46 hourly. Average hourly rates for field and livestock work have been on the rise since 1990. Language barriers, fear of deportation, frequent re-locations, and lack of voting status have contributed to difficulties in organizing farm laborers to advocate for wage, benefit, and working condition reforms.

The agricultural industry's reliance on non-native workers has become part of the political debate over immigration policies and enforcement in the country. United States Department of Labor statistics from 2009 indicated that about 50 percent of hired crop workers were not legally authorized to work in the U.S., a figure that remained largely unchanged over the course of the prior decade. During that same time period, intense debate took place in the nation's capital regarding immigration policies and enforcement. Farm interests, concerned about access to a steady workforce, worked on Capitol Hill to secure their interests in the Border Security, Economic Opportunity, and Immigration Modernization Act of 2013. Provisions favored by the farm lobby included: "earned adjustments" that will allow for temporary legal immigration status based on past experience with the possibility of applying for permanent residency by continuing to work in agriculture for a set period of time; and a more flexible guest worker program for agricultural workers.

In addition to farm labor, workers in the nation's food service industry garnered attention in 2012 and 2013 with a series of strikes against fast food outlets demanding higher wages, better working conditions, and the right to form unions. A study by the Labor Center at the University of California, Berkeley in October 2013 demonstrated that 52 percent of families of fast-food workers receive public assistance, in comparison to 25 percent of the workforce as a whole. According to the study, full-time hours were not sufficient to compensate for low wages.

Security

In the past, the denial of food deliveries has been used as a weapon in war. For example, during both world wars, the British naval blockade was intended to starve Germany into submission.

Food security is an important political issue as national leaders attempt to maintain control of sufficient food supplies for their nation. It can drive national policy, encourage the use of subsidies to stimulate farming, or even lead to conflict. This is mostly a national policy because nations have only recognized that there is a negative duty to not disrupt other nations food supply and do not require that one help them attain such secure access by protecting them against other threats.

In 1974, the World Food Summit defined food security as:

> *availability at all times of adequate world food supplies of basic foodstuffs to sustain a steady expansion of food consumption and to offset fluctuations in production and prices*

Food Security & Food Sovereignty in MENA Politics

"There are three traditional routes to national food security: 1) domestic production, which contributes to self-sufficiency; 2) commercial food imports; and 3) international food aid". Therefore, we need to make it clear that there is a distinction "between self-sufficiency and food security, in that the former is only one possible route to food security at the national level". Since 2007/2008, multiple Middle Eastern and North African (MENA) region governments have begun to consider more domestic food production as a part of their national aggregate food security laws. Although from a political view that approach may be justified due to it helping in stabilizing domestic food prices and deducting vulnerability to international markets and reliance on other countries, it comes at an enormous economic cost. This due to the resource endowments of the majority "of MENA countries—water scarcity and lack of arable land—are not well-suited to food production", specifically cereal production, and "these countries' comparative international advantages lie in other economic activities". Many of "the international organizations involved in the MENA economies during the 1990s and the 2000s advocated a food security strategy for most countries" that based on diversificating away from agriculture towards multiple other activities, "including manufacturing exports, with the resulting foreign exchange used to purchase food imports". "Within the agricultural sector there has also been emphasis on shifting resources into high- value crops that are most efficient in water use, such as fruits, vegetables, and tree crops", with a view on export markets, in replace of cereal production for domestic consumption.

Hunger

Malnutrition and starvation continue to be a persistent problem in some areas of the world. The effects of low agricultural output can be exacerbated by internecine struggles, such as the famine conditions that occurred in Somalia during the 1990s. But even under more stable conditions, hunger persists in some nations. Images of starvation can have a powerful influence, leading to charitable and even military intervention.

Retailing

During the late 1990s and early 21st century a significant amount of discussion and debate has developed surrounding the role of supermarkets in the retailing of food and the impacts of supermarkets both on the supply and production of food. Due to the buying power of the large supermarket chains, they can put huge demands on producers, often pushing prices artificially low, whilst still making large profits on the food themselves, with some products selling at over 400% the price paid to, whilst farmers may only make 50p profit on each animal produced domestically. This buying power also allows supermarkets to transcend national boundaries in sourcing food. For example, in the UK, where the food market is highly dominated by supermarkets, only 25% of apples sold in supermarkets are produced domestically, with out-of-season Cox apples being flown 14,000 miles from New Zealand, despite the UK being a natural producer of apples. Furthermore, due to the national nature of the supply networks used by supermarkets, this often involves domestically produced foodstuffs being transported around the country before being delivered to retailers, creating a huge impact both on traffic and pollution.

Climate Movement

The climate movement is the collective of nongovernmental organizations engaged in activism related to the issues of climate change. It is a subset of the broader environmental movement, but some regard it as a new social movement itself given its scope, strength and activities.

History

The climate movement has rapidly evolved in the first decades of the 21th century, starting as one of the many causes of the environmental movement.

Activism related to climate change began in the 1990s, when major environmental organizations became involved in the discussions about climate, mainly in the UNFCCC framework. In the 2000s several climate-specific organizations were founded, such as 350.org, Energy Action Coalition, and the Global Call for Climate Action.

Mobilization for Copenhagen 2009

The 2009 United Nations Climate Change Conference in Copenhagen was the first UNFCCC summit in which the climate movement started showing its mobilization power at a large scale. Between 40,000 and 100,000 people attended a march in Copenhagen on December 12 calling for a global agreement on climate. And activism went beyond Copenhagen, with more than 5,400 rallies and demonstrations took place around the world simultaneously.

Activities

The 2014 People's Climate March

The climate movement convened its largest single event in September 21, 2014, when it mobilized 400.000 activists in New York during the People's Climate March (plus several thousands more in other cities), to demand a climate action from the global leaders gathered for the 2014 UN Climate Summit.

Fossil Fuel Divestment

The Fossil Fuel Divestment movement calls for colleges and universities, as well as cities, states, religious institutions, and pension funds to withdraw their investments from fossil fuel companies. 350.org launched their Go Fossil Free campaign in 2012 with Bill McKibben's "Do The Math" speaking tour around the country. 350.org explains that the reasoning behind this campaign is simple: "If it is wrong to wreck the climate, then it is wrong to profit from that wreckage." The Fossil Fuel Divestment movement escalated tactics in the spring of 2015, with 19 students arrested at Yale University, a week-long blockade of Harvard University's administrative buildings, a 70-student sit-in at Tulane University, and other arrestable actions.

Climate Mobilization

Since 2014, growing portions of the climate movement, especially in the United States have been organizing for an international economic response to climate change on the scale of the mobilization of the American home front during World War II, with the goal of rapidly slashing carbon emissions and transitioning to 100% clean energy faster than the free market is likely to allow. Throughout 2015 and 2016, The Climate Mobilization led grassroots campaigns in the U.S. for this scale of ambition, and in July 2016, activists succeeded in getting text adopted into the Democratic Party's national platform calling for WWII-scale climate mobilization. In August 2015, environmentalist Bill McKibben published an article in the New Republic rallying Americans to "declare war on climate change."

Roles of other Movements

The climate movement is closely connected to other parts of the environmental movement, in particular groups aiming for a sustainable society and sustainable enery. Also the faith community has been active in the climate movement, both at an interfaith level (such as in Our Voices) and at the specific level of each denomination (such as the Global Catholic Climate Movement).

Animal Rights Movement

The animal rights movement, sometimes called the animal liberation movement, animal personhood, or animal advocacy movement, is a social movement which seeks an end to the rigid moral and legal distinction drawn between human and non-human animals, an end to the status of animals as property, and an end to their use in the research, food, clothing, and entertainment industries.

It is one of the few examples of a social movement that was created, and is to a large extent sustained academically, by philosophers.

Terms and Factions

All animal liberationists believe that the individual interests of non-human animals deserve recognition and protection, but the movement can be split into two broad camps.

Animal rights advocates, or rights liberationists, believe that these basic interests confer moral rights of some kind on the animals, and/or ought to confer legal rights on them; see, for example, the work of Tom Regan. Utilitarian liberationists, on the other hand, do not believe that animals possess moral rights, but argue, on utilitarian grounds — utilitarianism in its simplest form advocating that we base moral decisions on the greatest happiness of the greatest number — that, because animals have the ability to suffer, their suffering must be taken into account in any moral philosophy. To exclude animals from that consideration, they argue, is a form of discrimination that they call speciesism.

Despite these differences, the terms "animal liberation" and "animal rights" are generally used interchangeably.

Factional division has also been characterized as that between the reformist or mainstream faction and the radical abolitionist and direct action factions. The mainstream faction is largely professionalized and focuses on soliciting donations and gaining media representation. This focus on resource mobilization, sought by means of frame alignment with dominant cultural frames, leads to compromise with targets. Actors in the reformist movement diagnose the problem as one of nonhuman animal mistreatment, offer a prognosis including welfare reform, and employ action frames that include moral shocks. It has been noted that the power of the animal rights movement in the United States is centralized in professionalized nonprofit organizations that prefer moderated goals of welfare reform and tend to be isomorphic.

The abolitionist faction diagnoses the problem as one of nonhuman animal use, offers a prognosis including vegan outreach, and employ action frames including a rational and theoretic focus. Gary Francione, a leader in abolitionism, formed his approach in response to the traditional movement's focus on policy reform. Members of the abolitionist faction view policy reform as counterproductive and rely on nonviolent education and moral suasion in their repertoire of contention. They see the promotion of veganism as a low-opportunity cost means of creating an antispeciesist culture and posing an economic threat to the animal agriculture industrial complex.

The direct action or militant faction includes in its repertoire of contention property damage, open releases, intimidation, and direct violence. With these, they offer a prognosis to the problem of animal use that focuses on changing society through force and fear. Animal rights actors often reject this faction, pointing to violence as a counterproductive tactic that invites repression (e.g., the Animal Enterprise Terrorism Act) and does not economically or politically challenge extant systems.

Smaller factions include groups focused around faith-based animal rights theory and veganarchists, whose approach is characterized by a critique of capitalism on the grounds that it has led to mass nonhuman, human, and environmental exploitation.

Such factionalizing, researchers have pointed out, is common to social movements and plays a role in sustaining their health.

History

The Animal Rights Movement traces back to the animal protection movement in Victorian England, which was initiated by aristocratic moral crusaders in response to the poor treatment of urban workhorses and stray dogs. Other early influences include: Upton Sinclair's 1906 novel *The Jungle*, which drew attention to obscured slaughterhouse operations; Henry Salt's treatises on nonhuman animal rights; which drew from human abolitionist arguments for recognizing personhood of people considered to be property; and the Alcott House of New England, a community serving as a stop on the Underground Railroad and requiring its residents to eat a vegan diet.

Philosopher Peter Singer

The contemporary movement is regarded as having been founded in the UK in the early 1970s by a group of Oxford university post-graduate philosophy students, now known as the "Oxford Group". The group was led by Rosalind and Stanley Godlovitch, graduate students of philosophy who had recently become vegetarians. The Godlovitches met John Harris and David Wood, also philosophy graduates, who were soon persuaded of the arguments in favour of animal rights and themselves became vegetarian. The group began to actively raise the issue with pre-eminent Oxford moral philosophers, including Professor Richard Hare, both personally and in lectures. Their approach was based not on sentimentality ("kindness to dumb animals'), but on the moral rights of animals. They soon developed (and borrowed) a range of powerful arguments in support of their views, so that Oxford clinical psychologist Richard Ryder, who was shortly to become part of the group, writes that "rarely has a cause been so rationally argued and so intellectually well armed."

It was a 1965 article by novelist Brigid Brophy in *The Sunday Times* which was pivotal in helping to spark the movement. Brophy wrote:

> “ *The relationship of Homo sapiens* (humans) *to the other animals is one of unremitting exploitation. We employ their work; we eat and wear them. We exploit them to serve our superstitions: whereas we used to sacrifice them to our gods and tear out their entrails in order to foresee the future, we now sacrifice them to science, and experiment on their entrail in the hope — or on the mere offchance — that we might thereby see a little more clearly into the present.* ”

The philosophers found this article and were inspired by its vigorous unsentimental polemic. At about the same time, Ryder wrote three letters to the *Daily Telegraph* in response to Brophy's arguments. Brophy read Ryder's letters and put him in touch with the Godlovitches and John Harris, who had begun to plan a book about the issue which was also partly inspired by Brophy's polemic. The philosophers had also been to see Brophy about the possibility of a book of essays on the subject. They initially thought that a book with contributions from Brophy, Ruth Harrison, Maureen Duffy and other well-known writers might be of interest to publishers, but after an initial proposal was turned down by the first publisher they approached, Giles Gordon of Victor Gollancz suggested that the work would be more viable if it included their own writing. This was the idea that became “Animals, Men and Morals'.

Philosopher Tom Regan

In 1970, Ryder coined the phrase “speciesism,” first using it in a privately printed pamphlet to describe the assignment of value to the interests of beings on the basis of their membership of a particular species. Ryder subsequently became a contributor to *Animals, Men and Morals: An Inquiry into the Maltreatment of Non-humans* (1972), edited by John Harris and the Godlovitches, a work that became highly influential, as did Rosalind Godlovitch's essay “Animal and Morals,” published the same year.

It was in a review of *Animals, Men and Morals* for the *New York Review of Books* that Australian philosopher Peter Singer first put forward his basic arguments, based on utilitarianism and drawing an explicit comparison between women's liberation and animal liberation. Out of the review came Singer's *Animal Liberation*, published in 1975, now regarded by many as the “bible” of the movement.

Other books regarded as important include philosopher Tom Regan's *The Case for Animal Rights*

(1983); *Created from Animals: The Moral Implications of Darwinism* by James Rachels (1990); *Animals, Property, and the Law* (1995) by legal scholar Gary Francione, *Rattling the Cage: Toward Legal Rights for Animals* by another legal scholar Steven M. Wise (2000); and *Animal Rights and Moral Philosophy* by Julian H. Franklin (2005).

Gender, Class, and other Factors

The movement predominately comprises upper-class and middle-class white female members, owing this to its associations with the Victorian English animal protection movement and American feminism and environmentalism movements. As such, the movement is widely associated in public spheres with women, femininity, and effeminacy. Public perception of the movement is influenced by gendered evaluations; movement outsiders tend to view activists as irrational by virtue of overly emotional sentiments. Aware of this, activists have strategically incorporated men into positions of leadership and theory production, in order to legitimate the movement and counter popular beliefs about the primacy of emotion in the animal rights movement. This tactic relies on the popular perception of men as rational and not given to emotion, and follows a trend in social movement activism that seeks to counter traditional associations with femininity and private spheres by emphasizing rationality, rights, and justice. In one case study, targets of anti-hunting activism used class and gender markers to evaluate activists' claims. Hunters' associations of irrationality with femininity and of inexperience in hunting and wilderness with white-collar positions constituted the reasons for their dismissal of activists' claims. In contrast, hunters framed hunting in logical, scientific, and altruistic terms, thus legitimating hunting, termed wildlife management, as a protective measure.

It has been noted that the composition of the movement may discourage the mobilization of particular demographics. A content analysis of magazine covers from highly visible animal rights organizations (PeTA and VegNews) revealed that most featured members were white, female, and thin. With this, and with the composition of the movement being mostly white, female, and thin, it has been suggested that animal rights media depict an activist ideal-type with such characteristics, and that this may mobilize thin white females while deterring others. Racialized, sexualized, and size-focused campaign tactics may also serve to deter potential members from joining the movement. Racialized tactics include the appropriation of African slavery and Holocaust language and imagery, and have been deemed insensitive and impugned by nonwhite communities. In addition, the movement has maintained racist stereotypes about nonwhite individuals' predisposition toward animal cruelty; these stereotypes arose in post-slavery U.S. and Britain, where nonwhites were deemed by law and by society to have a tendency toward animal cruelty. Sexualization of "ideal" women is used as a mobilization tactic, but reduces support for ethics-based campaigns and may be counterproductive, alienating women that do not have "ideal" body types. Sizeism is used as a tactic to frame veganism as a healthy and positive lifestyle, aligning with a popular association of fatness with moral failure. These tactics may contribute to gender inequality because unrealistic and sexualized representations of women are linked to their societal devaluation. Its lack of diverse membership may decrease the movement's legitimacy and ability to mobilize, as members of marginalized groups are more likely to mobilize when they are represented in the movement. An inclusive movement with strong group solidarity would decrease opportunity costs associated with participating (e.g., social stigmatization, lack of alternatives, legal persecution) and thus serve to increase and sustain participation in the movement.

Current Status of the Movement

The movement is no longer viewed as hovering on the fringe. In the 1980s and 1990s, it was joined by a wide variety of academics and professionals, including lawyers, physicians, psychologists, veterinarians, and former vivisectionists, and is now a common subject of study in philosophy departments in Europe and North America. Animal law courses are taught in 92 out of 180 law schools in the U.S., and the movement has gained the support of senior legal scholars, including Alan Dershowitz and Laurence Tribe of Harvard Law School. Chapters of animal rights law have been created in several state bar associations, and resolutions related to animal rights are regularly proposed within the American Bar Association.

Michael Socarras of Greenberg Traurig told the Association of American Medical Colleges: "There is a very important shift under way in the manner in which many people in law schools and in the legal profession think about animals. This shift has not yet reached popular opinion. However, in [the U.S.], social change has and can occur through the courts, which in many instances do not operate as democratic institutions. Therefore, the evolution in elite legal opinion is extremely significant"

Philosophical and Legal Aims

The movement aims to include animals in the moral community by putting the basic interests of non-human animals on an equal footing with the basic interests of human beings. A basic interest would be, for example, not being made to suffer pain on behalf of other individual human or non-human animals. The aim is to remove animals from the sphere of property and to award them personhood; that is, to see them awarded legal rights to protect their basic interests.

> “Who are we that we have set ourselves up on this pedestal and we believe that we have a right to take from others everything—including their life—simply because we want to do it? Shouldn't we stop and think for a second that maybe they are just others like us? Other nations, other individuals, other cultures. Just others. Not sub-human, but just different from being human.”

Liberationists argue that animals appear to have value in law only in relation to their usefulness or benefit to their owners, and are awarded no intrinsic value whatsoever. In the United States, for example, state and federal laws formulate the rules for the treatment of animals in terms of their status as property. Liberationists point out that Texas Animal Cruelty Laws apply only to pets living under the custody of human beings and exclude birds, deer, rabbits, squirrels, and other wild animals not owned by humans, ignoring that jurisdiction for such creatures comes under the domain of state wildlife officers. The U.S. Animal Welfare Act excludes "pet stores ... state and country fairs, livestock shows, rodeos, purebred dog and cat shows, and any fairs or exhibitions intended to advance agricultural arts and sciences." There is no mention in the law that such activities already fall under the jurisdiction of state agriculture departments. The Department of Agriculture interprets the Act as also excluding cold-blooded animals, and warm-blooded animals not "used for research, teaching, testing, experimentation exhibition purposes, or as a pet, [and] farm animals used for food, fiber, or production purposes".

The Seattle-based Great Ape Project (GAP), founded by Peter Singer, is campaigning for the United Nations to adopt its Declaration on Great Apes, which would see chimpanzees, bonobos, gorillas and orang-utans included in a "community of equals" with human beings. The declaration wants to extend to the non-human apes the protection of three basic interests: the right to life, the protection of individual liberty, and the prohibition of torture.

Legal Changes Influenced by the Movement

Regarding the campaign to change the status of animals as property, the animal liberation movement has seen success in several countries. In 1992, Switzerland amended its constitution to recognize animals as *beings* and not *things*. However, in 1999 the Swiss constitution was completely rewritten. A decade later, Germany guaranteed rights to animals in a 2002 amendment to its constitution, becoming the first European Union member to do so. The German Civil Code had been amended correspondingly in 1997. The amendment, however, has not had much impact in German legal practice yet.

Perhaps the greatest success of the animal liberation movement has been the granting of basic rights to five great ape species in New Zealand in 1999. Their use is now forbidden in research, testing or teaching. Other governments had also previously implemented a ban on these experiments, such as the UK government in 1986. Some other countries have also banned or severely restricted the use of non-human great apes in research. Also, on May 17, 2013, India declared that all cetaceans have the status of "nonhuman persons."

In the United States, there is an Animal Welfare Act that was produced in 1966. This law protects animals in acts of research, transportation, and sale. Generally, animals are protected from any torture, neglect, or killing. There have been many amendments made towards this act to keep it updated. In order to be charged for animal abuse, there has to be prosecution and a trial. While there is only one act covering the entire United States, there are more current laws surrounding animal rights, which vary by state. Even though there are policies and severe consequences, the laws are not normally enforced with as much force or timeliness as regular crime.

Strategy and Tactical Repertoires

Use of New Information Communication Technologies (ICTs)

New media, such as the Internet and email, have been used by Animal Rights Movement actors and countermovement actors in a variety of capacities. Radical factions in the movement rely on websites, blogs, podcasts, videos, and online forums to engage in vegan outreach and other mobilization efforts and build alliances, thus overcoming exclusion by dominant factions. The use of Internet has allowed the animal rights movement to spread transnationally. For example, the Istanbul Animal Rights Movement's theory and repertoires of contention draw from those of various countries that have spread through the use of Internet.The Internet is also used by activists to build community and avoid stigmatization, and may be a preferred means of activism for marginalized members, such as individuals who are fat.

In 2001, Stop Huntingdon Animal Cruelty (SHAC), an animal rights group founded in the UK with the aim of ending vivisection practices by Huntingdon Life Sciences (HLS), published the names of targets associated with HLS on its website, situating the campaign against HLS in a globally accessible central information source that allowed for increased support and participation by decreasing opportunity cost. SHAC.net listed targets for "naming and shaming," emphasized and sent email action alerts, and facilitated written and digital communication between activists and targets. When the UK government later prevented SHAC from publishing reports from the ALF on its website, an activist created the Bite Back website, which was registered in the US and thus allowed the ALF to publish reports without reprisal. Countermovement actors have also used ICTs; law enforcement officers have tracked SHAC activists and admitted electronic communications as evidence in criminal trials. Dylan Barr, who jammed email inboxes at Washington Mutual Bank with 5,000 emails, caused $5,000 in losses and was convicted for extortion.

ICTs have facilitated undercover surveillance efforts by activists who use video cameras, Internet, and television to collect and disseminate evidence of cruelty to nonhuman animals, in order to attract publicity to and mobilize support for the movement.

Undercover Surveillance

In 1981, Alex Pacheco exposed illicit research on Macaque monkeys in Edward Taub's laboratory in Silver Spring, Maryland. Police raided the research facility, and the raid was televised, attracting publicity to antivivisectionists' claims. In the UK in 1990, Mike Huskisson and Melody McDonald videotaped Wilhelm Feldberg performing illegal research; the video evidence was made public and Feldberg's lab was summarily closed down. SHAC was founded after Zoe Broughton conducted undercover surveillance of vivisectionists and discovered evidence of nonhuman animal abuse. Footage and images from undercover surveillance activity are often circulated offline and on the Internet and used to deliver a moral shock that will mobilize viewers to participate in the movement. Members in the abolitionist faction, specifically those in Francione's camp, argue that the graphic depictions of suffering discovered in undercover work result in a focus on treatment, as opposed to use, and that this focus, while useful in securing welfare reform, is counterproductive to abolishing animal exploitation.

Boycott

Animal liberationists usually boycott industries that use animals. Foremost among these is factory farming, which produces the majority of meat, dairy products, and eggs in industrialized nations. The transportation of farm animals for slaughter, which often involves their live export, has in recent years been a major issue for animal rights groups, particularly in the UK and Scandinavia.

The vast majority of animal rights advocates adopt vegetarian or vegan diets. They may also avoid clothes made of animal skins, such as leather shoes, and will not use products known to contain animal byproducts. Goods containing ingredients that have been tested on animals are also avoided where possible. Company-wide boycotts are common. The Procter & Gamble corporation, for example, tests many of its products on animals, leading many animal rights advocates to boycott the company's products entirely, whether tested on animals or not.

There is a growing trend in the American movement towards devoting all resources to vegetarian outreach. The 9.8 billion animals killed there for food every year far exceeds the number of animals used in other ways. Groups such as Vegan Outreach and Compassion Over Killing devote their time to exposing factory-farming practices by publishing information for consumers and by organizing undercover investigations.

Moral Shocks

Moral shock is a tactic that involves drawing targets' attention to a particular depiction of a situation in order to cause outrage and catalyze targets to support a movement or claim. In the Animal Rights Movement, moral shocks are often used in the form of graphic depictions that detail the brutalization of nonhuman animals. Farm Animal Rights Movement (FARM), a popular animal rights organization, has used moral shocks in its pay-per-view campaign, in which passersby were paid a dollar to watch a graphic video of nonhuman animal suffering. Nonhuman animals depicted in moral shocks often display characteristics similar to those of human infants (e.g., large heads and eyes, crying or whimpering, small, mammalian). There is an ongoing debate within the Movement about the effectiveness of moral shocks. It has been found that many animal rights activists join after being exposed to moral shocks, and that moral shocks given to strangers are more likely to mobilize potential participants than are preexisting social networks; there is research that has found the opposite, however. Conversely, moral shocks that target the public at large (e.g., those used in vegan outreach) are less likely to be effective than those that have targets more distant from and less visible to the public (e.g., vivisectors).

Direct Action

A fire, claimed by the Oxford Arson Squad, caused £500,000 damage to Londbridges boathouse, Oxfordshire on July 4, 2005.

The movement espouses a number of approaches, and is bitterly divided on the issue of direct action and violence, with very few activists or writers publicly advocating the latter tactic as a justified method to use. Most groups reject violence against persons, intimidation, threats, and the destruction of property: for example, the British Union for the Abolition of Vivisection (BUAV) and Animal Aid. These groups concentrate on education and research, including carrying out undercover investigations of animal-testing facilities. There is some evidence of cooperation between the BUAV and the ALF: for example, the BUAV used to donate office space for the use of the ALF in London in the early 1980s.

Other groups do not condemn the destruction of property, or intimidation, but do not themselves engage in those activities, concentrating instead on education, research, media campaigns, and undercover investigations: for example, People for the Ethical Treatment of Animals (PETA).

A third category of activists operates using the leaderless resistance model, working in covert cells consisting of small numbers of trusted friends, or of one individual acting alone. These cells engage in direct action: for example by carrying out raids to release animals from laboratories and farms, using names like the Animal Liberation Front (ALF); or by boycotting and targeting anyone or any business associated with the controversial animal testing lab, Huntingdon Life Sciences (HLS), using a campaign name like Stop Huntingdon Animal Cruelty (SHAC). Some arson, property destruction and vandalism has been linked to various animal rights groups.

Activists who have carried out or threatened acts of physical violence have operated using the names; Animal Rights Militia (ARM), Justice Department, Revolutionary Cells—Animal Liberation brigade (RCALB), Hunt Retribution Squad (HRS) and the Militant Forces Against Huntingdon Life Sciences (MFAH).

Some activists have attempted blackmail and other illegal activities, such as the intimidation campaign to close Darley Oaks farm, which involved hate mail, malicious phonecalls, bomb threats, arson attacks and property destruction, climaxing in the theft of the corpse of Gladys Hammond, the owners' mother-in-law, from a Staffordshire grave. Over a thousand ALF attacks in one year in the UK alone caused £2.6M of damage to property, prompting some experts to state that animal rights now tops the list of causes that prompt violence in the UK.

There are also a growing number of "open rescues," in which liberationists enter businesses to remove animals without trying to hide their identities. Open rescues tend to be carried out by committed individuals willing to go to jail if prosecuted, but so far no farmer has been willing to press charges.

Many of the ideas used by those who engage in direct action were developed by British activists.

Targeting Researchers

Activists have targeted individual researchers and have shown up at homes in the middle of the night, threatening their families and children.

Overwhelmingly, though, the animal rights movement is peaceful, and instances of violence have been used in efforts to try to tarnish the entire movement.

Criminalization of Direct Action Methods

The U.S. Justice Department labels underground groups the Animal Liberation Front and the Earth Liberation Front as terrorist organizations.

A November 13, 2003 edition of CBS News' *60 Minutes* charged that "eco-terrorists," a term used by the United States government to refer to the Animal Liberation Front and Earth Liberation Front, are considered by the FBI to be "the country's biggest domestic terrorist threat." John Lewis, a Deputy Assistant Director for Counterterrorism at the FBI, stated in a 60 Minutes interview that these groups "have caused over $100 million worth of damage nationwide", and that "there are more than 150 investigations of eco-terrorist crimes underway".

"Animal Enterprise Terrorism Act", the legislation which allows federal authorities to "help prevent, better investigate, and prosecute individuals who seek to halt biomedical research through acts of intimidation, harassment, and violence" was adopted in the USA in 2006. It has also been described as having 'a chilling effect' on free speech.

Inter-Movement Activity

Radical animal rights factions address injustices against a variety of groups, therein emphasizing a connection between discrimination against humans and discrimination against nonhuman animals. An intersectional orientation is seen online, on websites and social media, and also in offline activity. In Turkey, animal rights groups commonly join other social movements by aligning with online and offline campaigns. In Istanbul's 2013 Gezi Park protests, which began as an environmental movement against urban development efforts, various social movement groups participated. Among them were animal rights activists that saw the protests as an opportunity to raise concerns about speciesism. Animal rights activists' involvement in the protests changed the opinions of animal rights movement outsiders who had previously viewed vegan animal rights activists as elitist. This allowed for increased legitimacy and network expansion; the animal rights movement in Istanbul is composed of multi-movement actors from the feminist movement, LGBT+ movement, and antimilitarist movement, and such inter-movement interaction has led to increased coverage of veganism and animal rights by leftist news sites in Turkey.

Countermovement

Opposition to the Animal Rights Movement comes from corporate and state actors. Mass media, agribusiness, and biomedicine industries often portray activists in a negative light, characterizing the movement as misanthropic, sensationalist, and dangerous to scientific endeavors and human wellbeing because of activists' high levels of expressed empathy for nonhuman animals. Mass media also frequently portray nonhuman animals as objects. Major pharmaceutical companies have taken legal measures to disallow protestors from targeting their companies.

The abolitionist faction of the Animal Rights Movement often faces counterframing by dominant reformist organizations of the movement that frame radical advocacy as idealistic and schismatic. These reform-oriented nonhuman rights organizations direct resources to countering abolitionist claims and blocking abolitionists' access to discursive spheres. Another example of counterframing from opposition movement actors is found in Switzerland's 1998 referendum cycle, in which antivivisectionists' claims that animal research should be abolished were contested with claims that mobilized the public more. Antivivisection claims, which framed animal research as facilitating the genetic engineering of foods in hopes of seizing on public fear of genetic engineering, were countered by scientists and animal researchers, who framed vivisection as medically necessary to ensure human wellbeing.

References

- Messer, E; Cohen, M (Summer 2007). "Conflict, Food Insecurity, and Globalization". Food, Culture and Society: An International Journal of Multidisciplinary Research. 2. 10: 297–315. doi:10.2752/155280107x211458
- Barton, Greg (2002). Empire Forestry and the Origins of Environmentalism. Cambridge University Press. p. 48. ISBN 9781139434607
- Griswold, Eliza (21 September 2012). "How 'Silent Spring' Ignited the Environmental Movement". New York Times. Retrieved 22 February 2013

- Djoundourian, Salpie (2011-08-01). "Environmental movement in the Arab world". Environment, Development and Sustainability. 13 (4): 743–758. doi:10.1007/s10668-011-9287-7
- Stuckler, D; Nestle, M (2012). "Big food, big systems, and global health". PLoS Medicine. 96 (6): e1001242. doi:10.1371/journal.pmed.1001242
- Hunt, Michael H. (2004). The World Transformed: 1945 to the present. New York, New York: Oxford University Press. pp. 215–217. ISBN 9780199371020
- Brooke, Elizabeth Heilman (2 June 1992). "As Forests Fall, Environmental Movement Rises in Brazil". The New York Times. Retrieved 20 February 2013
- Wallinga, D; Burch, D (July 10, 2013). "Head to Head: Does adding routine antibiotics to animal feed pose a serious risk to human health?". BMJ. 347: f4214. doi:10.1136/bmj.f4214
- Greenwood, Jeremy; Seshadri, Ananth; Yorukoglu, Mehmet (January 2005). "Engines of liberation". The Quarterly Journal of Economics. Oxford Journals. 72 (1): 109–133. doi:10.1111/0034-6527.00326. Pdf
- Wu and Wen (2015). Nongovernmental organizations and environmental protests: Impacts in East Asia (chapter 7 of Routledge Handbook of Environment and Society in Asia). London: Routledge. pp. 105–119. ISBN 978-0-415-65985-7
- Bello, Walden. "The Environmental Movement in the Global South". Transnational Institute. Environmental Justice. Retrieved 15 February 2013
- Ifland, JR; et al. (May 2009). "Refined Food addiction: A classic substance disorder". Medical Hypothesis. 72 (5): 518–526. doi:10.1016/j.mehy.2008.11.035
- Aguiar, Mark; Hurst, Erik (August 2007). "Measuring trends in leisure: the allocation of time over five decades". The Quarterly Journal of Economics. Oxford Journals. 122 (3): 969–1006. doi:10.1162/qjec.122.3.969. Pdf
- Directorate-General for Research and Innovation: Biotechnologies, Agriculture, Food. European Union 2010. "A decade of EU-funded GMO research (2001–2010)." doj: 10.2777/97784. ISBN 978-92-79-16344-9
- "Report 2 of the Council on Science and Public Health: Labeling Bioengineered Foods" (PDF). American Medical Association. Retrieved 2013-11-15
- Laestadium, LI; Neff, RA; Barry, CL; Frattaroli, S (2013). "Meat consumption and climate change: the role of non-governmental organizations". Climatic Change. 120: 25–38. doi:10.1007/s10584-013-0807-3
- "University of Saskatchewan Policies 2001: Gender Neutral Language". University of Saskatchewan. Archived from the original on October 28, 2006. Retrieved March 25, 2007

Permissions

All chapters in this book are published with permission under the Creative Commons Attribution Share Alike License or equivalent. Every chapter published in this book has been scrutinized by our experts. Their significance has been extensively debated. The topics covered herein carry significant information for a comprehensive understanding. They may even be implemented as practical applications or may be referred to as a beginning point for further studies.

We would like to thank the editorial team for lending their expertise to make the book truly unique. They have played a crucial role in the development of this book. Without their invaluable contributions this book wouldn't have been possible. They have made vital efforts to compile up to date information on the varied aspects of this subject to make this book a valuable addition to the collection of many professionals and students.

This book was conceptualized with the vision of imparting up-to-date and integrated information in this field. To ensure the same, a matchless editorial board was set up. Every individual on the board went through rigorous rounds of assessment to prove their worth. After which they invested a large part of their time researching and compiling the most relevant data for our readers.

The editorial board has been involved in producing this book since its inception. They have spent rigorous hours researching and exploring the diverse topics which have resulted in the successful publishing of this book. They have passed on their knowledge of decades through this book. To expedite this challenging task, the publisher supported the team at every step. A small team of assistant editors was also appointed to further simplify the editing procedure and attain best results for the readers.

Apart from the editorial board, the designing team has also invested a significant amount of their time in understanding the subject and creating the most relevant covers. They scrutinized every image to scout for the most suitable representation of the subject and create an appropriate cover for the book.

The publishing team has been an ardent support to the editorial, designing and production team. Their endless efforts to recruit the best for this project, has resulted in the accomplishment of this book. They are a veteran in the field of academics and their pool of knowledge is as vast as their experience in printing. Their expertise and guidance has proved useful at every step. Their uncompromising quality standards have made this book an exceptional effort. Their encouragement from time to time has been an inspiration for everyone.

The publisher and the editorial board hope that this book will prove to be a valuable piece of knowledge for students, practitioners and scholars across the globe.

Index